The History of Mathematics: A Very Short Introduction

VERY SHORT INTRODUCTIONS are for anyone wanting a stimulating and accessible way into a new subject. They are written by experts, and have been translated into more than 45 different languages.

The series began in 1995, and now covers a wide variety of topics in every discipline. The VSI library now contains over 500 volumes—a Very Short Introduction to everything from Psychology and Philosophy of Science to American History and Relativity—and continues to grow in every subject area.

Titles in the series include the following:

Jacqueline Stedall

THE HISTORY OF MATHEMATICS

A Very Short Introduction

OXFORD
UNIVERSITY PRESS

Great Clarendon Street, Oxford OX2 6DP

Oxford University Press is a department of the University of Oxford.
It furthers the University's objective of excellence in research, scholarship,
and education by publishing worldwide in

Oxford New York

Auckland Cape Town Dar es Salaam Hong Kong Karachi
Kuala Lumpur Madrid Melbourne Mexico City Nairobi
New Delhi Shanghai Taipei Toronto

With offices in

Argentina Austria Brazil Chile Czech Republic France Greece
Guatemala Hungary Italy Japan Poland Portugal Singapore
South Korea Switzerland Thailand Turkey Ukraine Vietnam

Oxford is a registered trade mark of Oxford University Press
in the UK and in certain other countries

Published in the United States
by Oxford University Press Inc., New York

British Library Cataloguing in Publication Data

Data available

Library of Congress Cataloging in Publication Data

Data available

Typeset by SPI Publisher Services, Pondicherry, India
Printed in Great Britain on acid-free paper by
Ashford Colour Press Ltd., Gosport., Hampshire.

ISBN 978-0-19-959968-4

Contents

Contents

Acknowledgements

In writing this *Very Short Introduction* to a very large topic
I have been greatly inspired by other authors in the series, so
many of whom have risen to a similarly demanding challenge
in imaginative and thought-provoking ways.

It has been my privilege over the last few years to edit both *The
Oxford Handbook of the History of Mathematics* and the *BSHM
Bulletin*, the journal of the British Society for the History of
Mathematics. This has led me into close working relationships
with more than eighty authors writing on the history of
mathematics from a wide variety of perspectives. I have learned
something from every one of them. Much of that work was done
alongside Eleanor Robson, best of friends and colleagues, and I
am immensely grateful to her for the hours of companionship and
discussion that have helped to shape the picture I have attempted
to convey in this book. In particular, I have drawn on the research
and expertise of Markus Asper, Sonja Brentjes, Christopher
Cullen, Marit Hartveit, Annette Imhausen, Kim Plofker, Eleanor
Robson, Corinna Rossi, Simon Singh, Polly Thanailaki, and
Benjamin Wardhaugh; books and articles by these authors and
some others will be found in the suggestions for further reading at
the end.

The John Hersee collection of children's copy books, discussed in Chapter 4, is the property of the Mathematical Association, and is housed in the David Wilson Library at the University of Leicester. I thank the Association's archivists, Mary Walmsley and Mike Price, for their generous hospitality and collaboration in that part of my research. I am also grateful to Joanna Parker of Worcester College, Oxford, for allowing me to see John Aubrey's copy of Anne Ettrick's notebook. I am indebted to Andrew Wiles, Christopher Cullen, Eleanor Robson, and Adam Silverstein for taking the trouble to check details in Chapters 1, 2, 4, and 5 respectively. I record my warm thanks to them and all others who have commented astutely on various aspects of the text: anonymous OUP readers, together with Peter Neumann, Harvey Lederman, Jesse Wolfson, and all the members of my immediate family, some of whom never thought to read about the history of mathematics until now.

List of illustrations

Introduction

Mathematics has a history that stretches back for at least 4,000 years and reaches into every civilization and culture. It might be possible, even in an introduction as very short as this book, to outline some key mathematical events and discoveries in roughly chronological order. Indeed, this is probably what most readers will expect. There can be several problems, however, with that kind of exposition.

The first is that such accounts tend to portray a whig version of mathematical history, in which mathematical understanding is generally perceived to progress onwards and upwards towards the splendid achievements of the present day. Unfortunately, those looking for evidence of progress tend to overlook the complexities, lapses, and dead ends that are an inevitable part of any human endeavour, including mathematics; sometimes failure can be as revealing as success. Besides, by defining present-day mathematics as the benchmark against which earlier efforts are to be measured, we can too easily come to regard the contributions of the past as valiant but ultimately outdated efforts. Instead, in looking to see how this or that fact or theorem originated, we need to see discoveries in the context of their own time and place.

A second problem, about which I shall have more to say later, is that chronological accounts all too often follow a 'stepping stone' style, in which discoveries are placed before us one after another without the all-important connections between them. The aim of the historian is not merely to compile dated lists of events but to throw light on the influences and interactions that led to them. This will be a recurring theme of this book.

A third problem is that key events and discoveries come to be associated with key people. Further, in most histories of mathematics, most of those people will have lived in western Europe from about the 16th century onwards and will be male. This does not necessarily reflect Eurocentric or sexist attitudes on the part of the writers. The rapid pace of development of mathematics in the masculine culture of Europe since the Renaissance has led to a large amount of material that historians have rightly thought worth investigating; besides, we have a wealth of sources from Europe for this period, as opposed to only a handful, in relative terms, for pre-medieval Europe, China, India, or America. Fortunately, the availability and accessibility of sources from some of these other areas is beginning to improve. The fact remains, however, that focussing on big discoveries rules out the mathematical experience of most of the human race: women, children, accountants, teachers, engineers, factory workers, and so on, often entire continents and centuries of them. Clearly this will not do. Without denying the value of certain notable breakthroughs (and this book will begin with one of them) there have to be ways of thinking about history in terms of the many who practise mathematics, not just a few.

This book can do only a little to redress the masculine bias of most depictions of the history of mathematics; it can, however, pay more than lip service to the mathematics of continents other than Europe; and it will attempt to explore how, where, and why mathematics has been practised by people whose names will never appear in standard histories. To do so, however, requires something different from the usual chronological survey.

The alternative model that I propose to follow will be built around themes rather than periods. Each chapter will focus on two or three case studies, chosen not because they are in any way comprehensive or exhaustive but in the hope that they will suggest ideas and questions and fresh ways of thinking. At the same time, in keeping with the ideals proclaimed above, I have tried wherever possible to draw out contrasts or similarities between the various stories, so that readers can build up an interconnected view of at least a few aspects of the very long history of mathematics. My aim has been to demonstrate not only how professional historians now approach their discipline but how the layperson too can think about mathematical history.

In this way, I hope that this book will help the reader to recognize the richness and diversity of mathematical activity throughout human history; and that it will be a very short introduction not just to some of the mathematics of the past but to the history of mathematics itself as a modern academic discipline.

Chapter 1
Mathematics: myth and history

It is not often that a thorny old mathematical problem makes the news, but in 1993 newspapers in Britain, France, and the United States announced that a 40-year-old mathematician called Andrew Wiles, in a lecture at the Isaac Newton Institute in Cambridge, had demonstrated a proof of a 350-year-old problem known as Fermat's Last Theorem. As it turned out, the claim was a little premature: Wiles's 200 pages of mathematics contained an error that took a little while to fix, but two years later the proof was secure. The story of Wiles's nine-year struggle with the theorem became the subject of a book and of a television film in which Wiles was moved to tears as he spoke of his final breakthrough.

One reason that this piece of mathematical history so caught the public imagination was undoubtedly the figure of Wiles himself. For seven years before the Cambridge lecture, he had worked in near isolation, devoting himself single-mindedly to the deep and complicated mathematics underlying the theorem. Here then was a story to which those brought up in the mythologies of western culture were already well attuned: the lonely hero struggling against the odds to attain an elusive goal. There was even a princess in the background: only his wife knew of Wiles's ultimate purpose, and was the first to receive the finished proof, as a birthday present.

A second reason is that although the eventual proof of Fermat's Last Theorem was fully understood by perhaps no more than 20 people in the world, the theorem itself is easily stated. Wiles was already intrigued by it when he was 10 years old, and even those who have long ago forgotten most of the mathematics they ever learned can grasp what it is about; we will return to it in a moment.

Before that, however, note that in the very first sentence of this chapter three people were already mentioned by name: Wiles, Newton, and Fermat. In mathematics this is typical: it is universal practice for mathematicians to name theorems, conjectures, or buildings after one of the tribe. This is because most mathematicians are keenly aware that they continuously build on work done by their predecessors or their colleagues. In other words, mathematics is an innately historical subject in which past endeavours are rarely far out of mind. To begin to think about the questions that historians of mathematics ask, let us pursue Fermat's Last Theorem backwards from that Cambridge lecture theatre in 1993 to its more remote beginnings.

Fermat and his theorem

Pierre de Fermat, born in 1601, spent his entire life in southern France. A lawyer by training, he was a Counselor to the Parlement of Toulouse, the judicial body for a large surrounding area. In his spare time, which was little enough, Fermat worked on mathematics, and being far removed from circles of intellectual activity in Paris he did so almost entirely alone. During the 1630s he corresponded with mathematicians further afield, through the Parisian Minim friar Marin Mersenne, but during the 1640s, as the political pressures on him increased, he withdrew once again into mathematical isolation. Fermat achieved some of the most profound results of early 17th-century mathematics but for the most part was prepared to say tantalizingly little about them. Time and again he promised his correspondents that he would fill in the details when he had enough leisure to do so, but that leisure never came. Sometimes he would offer a bare statement of what he had found, or he would send out challenges that plainly demonstrated the ideas he was working on but without giving away his hard-won results.

The first hint of his Last Theorem appeared in such a challenge, sent to the English mathematicians John Wallis and William Brouncker in 1657; they failed to see what he was driving at and dismissed it as being beneath their dignity. Only after Fermat's death, when some of his notes and papers were edited by his son Samuel, did the full statement of the theorem emerge, scribbled in the margin of Fermat's copy of the *Arithmetica* of Diophantus. Before taking another step back in time to see what it was in Diophantus that inspired Fermat, we need to digress briefly to some mathematics, Fermat's Last Theorem itself.

The one bit of mathematics that almost everyone recalls from their schooldays is Pythagoras' Theorem, which states that the square on the longest side of a right-angled triangle, the hypotenuse, is equal to the sum of the squares on the two shorter sides, the 'legs'. Most people will probably also remember that if the two short sides are respectively 3 and 4 units in length then the long side will be 5 units, because $3^2 + 4^2 = 5^2$. This kind of triangle is known as a 3-4-5 triangle and may conveniently be used for marking out right angles on the ground with a piece of rope, or by textbook writers who want to set problems that can be solved without resort to a calculator. There are plenty of other sets of three whole numbers that satisfy the same relationship: it is easy to check that $5^2 + 12^2 = 13^2$, for instance, or that $8^2 + 15^2 = 17^2$. Such sets, sometimes written as (3, 4, 5), (5, 12, 13), and so on, are known as 'Pythagorean triples', and there are infinitely many of them.

Now suppose that, as mathematicians like to do, we tweak the conditions a bit and see what happens. What if, instead of taking the squares of each number, we take their cubes? Can we find

3

triples (a, b, c) that satisfy $a^3 + b^3 = c^3$? Or can we be even wilder and ask for a triple that satisfies $a^7 + b^7 = c^7$ or even $a^{101} + b^{101} = c^{101}$? The conclusion Fermat came to was that there is no point in trying: we cannot do it for any power beyond squares. As so often, however, he left it to others to work out the details. This time, his excuse was not time but space: he had discovered a marvellous proof, he said, but the margin was too meagre to contain it.

The margin in question belonged to page 85 of Claude Gaspar Bachet's 1621 edition of the *Arithmetica* of Diophantus. The *Arithmetica* had intrigued European mathematicians ever since a manuscript copy, written in Greek, had been rediscovered in Venice in 1462. About Diophantus himself, no-one knew anything, and little more is known now. The manuscript refers to him as 'Diophantus of Alexandria' so we may suppose that he lived and worked for a significant part of his life in that Greek-speaking city of northern Egypt. Whether he was a native Egyptian or an incomer from some other part of the Mediterranean world, we do not know. And any estimate for his dates is no more than guesswork. Diophantus cited a definition from Hypsicles (c. 150 BC), while Theon (c. AD 350) cited a result from Diophantus. That pins him down to within 500 years, but we cannot do better than that.

Compared with the geometric texts that have survived from other Greek mathematical writers, the *Arithmetica* is highly unusual. Its subject matter is not geometry but nor is it the arithmetic of everyday accounting. Rather, it is a set of sophisticated problems asking for whole numbers or fractions that must satisfy certain conditions. The eighth problem of the second book, for example, asks the reader to 'divide a square into two squares'. For our present purposes, we may translate this into a more modern mode of expression and see that Diophantus' question was related to Pythagorean triples, where a given square (in the notation above, c^2) may be divided or separated into two smaller squares ($a^2 + b^2$).

4

Diophantus showed a clever way of achieving this when the largest square is 16 (in which case the answer involves fractions); and then he moved on to something else.

Fermat, however, hesitated at this point and must have asked himself the obvious question: can the method be extended? Can one 'divide a cube into two cubes'? This was precisely the question he posed to Wallis and Brouncker in 1657 (and to which, after Fermat had later reported that it was impossible, Wallis angrily retorted that such 'negative' questions were absurd). What Fermat suggested in the margin in fact applied not only to cubes but to any higher power at all, a long way beyond anything required by Diophantus.

One other name has recurred throughout the above account, so let us now take one further historical step backwards, from Diophantus to Pythagoras, who is supposed to have lived on the Greek island of Samos around 500 BC. Despite this very early date, many readers will probably feel much more at home with Pythagoras than with Diophantus: indeed the question I am most commonly asked as a historian of mathematics is: 'Do you go all the way back to Pythagoras?' It is true that Pythagoras' Theorem has been known for a very long time; the disappointing news is that there is no evidence to link it to Pythagoras. In fact, there is little evidence to link anything to Pythagoras. If Diophantus is a shadowy figure, Pythagoras is buried under a blanket of myth and legend. We have no texts written by him or his immediate followers. The earliest surviving accounts of his life are from the third century AD, about 800 years after he lived, by writers with their own philosophical axes to grind. His supposed journeys to Babylon, or to Egypt, where he was said to have learned geometry, were probably no more than fictions used by such writers to bolster Pythagoras' standing and authority. As to the tales of what his followers are supposed to have done or believed, there may be some foundation for them in fact, but it is impossible to be certain of any of them. In short, Pythagoras became, literally, a legendary figure, to whom much was ascribed but of whom little was known in reality.

The lives of these four men, Pythagoras, Diophantus, Fermat, Wiles, span more than 2,000 years of mathematical history. We can certainly trace similar mathematical ideas running through the stories about each of them even though they are spaced several centuries apart. Have we then 'done' the history of Fermat's Last Theorem from start to finish? The answer is 'no', and for many reasons. The first is that one task of the historian is to disentangle fiction from fact, and myth from history. This is not to underestimate the value of either fiction or myth: both embody the stories by which societies define and understand themselves, and may have deep and lasting value. The historian, however, must not allow those stories to obscure evidence that may point to other interpretations. In the case of Pythagoras, it is relatively easy to see how and why tales that appear robust have been spun from the flimsiest of threads, but in the case of Andrew Wiles, where we believe we have the facts in front of our eyes, it is much harder. The truth of almost any story is almost always more complex than we first imagine or than the authors would sometimes have us believe, and stories about mathematics and mathematicians are no exception. The rest of this chapter examines some common myths and pitfalls in the history of mathematics; for convenience, I have called them 'Ivory tower history', 'Stepping-stone history', and 'Elite history'. The rest of the book will then offer some alternative approaches.

Ivory tower history

One of the most remarkable features of Wiles's story is the fact that he deliberately shut himself away for seven years so that he could pursue the proof of the Last Theorem without interruption or interference. Fermat too was clearly a loner, separated by geographical distance if nothing else from those who might have been able to understand and appreciate his work. We

have spoken of Diophantus and Pythagoras also without any
reference to their contemporaries. Were these four men then really
lonely geniuses forging new paths alone? Is this how mathematics
is done properly, or done best? Let us return to Pythagoras and
this time work forward.

The stories about Pythagoras persistently claim that he
established or attracted around him a community, or brotherhood,
who shared certain religious and philosophical beliefs and perhaps
also some mathematical explorations. Unfortunately the stories
also claim that the brotherhood was bound to strict secrecy, which
of course leaves room for endless speculation about their activities.
Even if there is only a grain of truth in such stories, however, it
would seem that Pythagoras was charismatic enough to attract
followers. Indeed, the fact that his name has survived at all
suggests that he was respected and revered in his lifetime, and
that he was no hermit.

We are a little better able to place Diophantus, who in Alexandria
would have been able to enjoy the company of other scholars. He
would also almost certainly have had access to books gathered
from other parts of the Mediterranean world in temple or private
book collections. It is possible that the problems of the
Arithmetica were his own invention, but it could equally be the
case that he compiled them into a single collection from a variety
of other sources, written or oral. One of the recurring motifs of
this book will be that mathematics repeatedly passes from one
person to another through the spoken word. Diophantus, like any
other mathematically creative person, almost certainly discussed
his problems and their solutions with a teacher or with students of
his own. We should therefore think of him not as a silent figure
writing his books in private but as a citizen of a city where learning
and intellectual exchange were valued.

Even Fermat, confined to Toulouse and the rigours of full-time
political employment, was not quite as isolated as might first

appear. One of his friends during his early studies in Bordeaux was Etienne d'Espagnet, whose father had been a friend of the French lawyer and mathematician François Viète. The works of Viète, otherwise rare but thus made available to Fermat, were to have a profound influence upon his development as a mathematician. Another friend, and fellow Counselor in Toulouse, was Pierre de Carcavi, who, when he moved to Paris in 1636, took with him news of Fermat and his discoveries. Through Carcavi, Fermat became known to Marin Mersenne, and through Mersenne he corresponded with Roberval, probably the best mathematician in Paris at the time, and with Descartes in the Netherlands. Later he communicated some of the discoveries arising from his studies of Diophantus to Blaise Pascal in Rouen and to John Wallis in Oxford. Thus even Fermat, far from important centres of learning, was connected into a network of correspondence that stretched across Europe, a virtual community of scholars that later came to be called the Republic of Letters.

When it comes to Wiles, it is much easier to see the cracks in the 'lone genius' story: Wiles was educated at Oxford and Cambridge, and later worked at Harvard, Bonn, Princeton, and Paris, in all of which he was part of flourishing mathematical communities. The mathematical clue that eventually gave direction to his interest in the Last Theorem was picked up from a casual conversation with a fellow mathematician in Princeton; when after five years he needed a fresh breakthrough, he attended an international conference in order to elicit the latest thinking on the subject; when he needed technical help with an important aspect of the proof, he broke his secrecy to a colleague, Nick Katz, and delivered the material in question in a graduate lecture course, though it eventually lost all its listeners except Katz; two weeks before he made the entire proof public in three lectures in Cambridge, England, he asked a colleague, Barry Mazur, to check it; the final proof was checked by six others; and when a flaw was discovered, Wiles invited one of his former students, Richard Taylor, to help him fix it. Further, throughout his years of working on the proof,

Wiles never stopped teaching students or attending departmental seminars. In short, although he spent many hours alone, he was also embedded in a community that allowed him to do so, and which, when required, came to his aid.

Wiles's years of isolation capture the imagination not because they are normal for a working mathematician but because they were exceptional. Mathematics is fundamentally and necessarily a social activity at every level. Every mathematics department in the world contains communal spaces, whether alcoves or common rooms, always equipped with some kind of writing surface, so that mathematicians can put their heads together over the tea and coffee that fuel them. Language or history students rarely write their essays collaboratively, and would not be encouraged to do so, but mathematics students frequently and fruitfully work together, teaching and learning from each other. And despite all the advances of modern technology, mathematics is still primarily learned not so much from books as from other people, through lectures, seminars, and classes.

Stepping stone history

In the outline of the story of Fermat's Last Theorem sketched above, Pythagoras, Diophantus, Fermat, and Wiles appear not only as isolated in their own lives but also from one other, like stepping stones standing out across an otherwise featureless river. If the ivory tower version of history isolates mathematicians from their social groups and communities, the stepping stone version isolates them from their past. Since the past is supposed to be the subject of history, it seems strange to ignore huge chunks of it in this way, but a surprising number of general histories of mathematics are presented in stepping stone style.

Let us then re-examine our story and the gaps in it a little more closely. Just as Pythagoras and Diophantus are somewhat shadowy, so is the space between them. It is possible that

Diophantus had never heard of Pythagoras. He would almost certainly, however, have come across 'Pythagoras' Theorem', not from any writings of Pythagoras, but in the work of Euclid, who lived around 250 BC. Apart from this very approximate date, we know no more about Euclid than about Diophantus a few centuries later, but his master work, the *Elements*, survived to become the longest-running textbook ever, still used in school geometry teaching well into the 20th century. The *Elements* is a comprehensive compilation of the geometry of Euclid's day, with the theorems arranged in careful logical order, and the penultimate theorem of the first book is 'Pythagoras' Theorem', carefully proved by geometric construction. One may reasonably suppose that Diophantus in Alexandria had access to the *Elements*, and it is possible that 'Pythagoras' Theorem' set him thinking about Pythagorean triples. It is equally possible, however, that his inspiration came from other sources that we no longer know about.

The first few centuries between Diophantus and Fermat are almost harder to fill in than those before Diophantus, even in the imagination. We know that Diophantus' *Arithmetica* was originally written in thirteen books, but only the first six survived in Greek; we do not know how or why. (In 1968 an Arabic manuscript was discovered in Iran which claims to be a translation of books IV to VII, but scholars are not agreed as to how accurately the text represents the original.) Fortunately, those six books were preserved for the Greek-speaking world at Byzantium (later Constantinople, now Istanbul), and eventually copies were brought to western Europe. As will be discussed further in Chapter 6, a German scholar known as Regiomontanus saw one of them in Venice in 1462 and believed that it contained the origins of the outlandish subject known to Europeans as algebra. A century later, the Italian engineer and algebraist Rafael Bombelli studied a manuscript of the *Arithmetica* in the Vatican and halted work on his own book on algebra in order to incorporate problems from Diophantus. The first printed edition was published in Basle in 1575, in Latin, translated and edited by Wilhelm Holtzman (Xylander), a humanist scholar, who described the work as 'incomparable, containing the true perfection of arithmetic'. The problems of Diophantus continued to intrigue those who came across them, and in 1621 a new Latin edition of the *Arithmetica* was produced by Claude Gaspard Bachet de Méziriac in Paris. This was the edition that Fermat owned and annotated.

It is not too difficult to fill in the gap between Fermat and Wiles. The Last Theorem, published by Samuel Fermat in 1670, seems not to have attracted any serious attempts in the 17th century, but in the 18th century it came to the attention of Leonhard Euler, the most versatile and prolific mathematician of the period, who made some inroads into the easier cases of it. In 1816, the Paris Academy of Sciences offered a prize for a solution. This inspired the efforts of Sophie Germain, who had some success with certain parts of it and whose work was taken up and extended by others. Beyond that, the problem gradually became widely known and over the years attracted hundreds, if not thousands, of purported solutions, from professionals and amateurs alike. Most of these attempts were both incorrect and useless, but a few led to important mathematical discoveries in their own right, which Wiles would have known about. When he eventually embarked on his own proof, he used some of the deepest mathematics of the 20th century, which was by then known to relate to Fermat's Last Theorem: the Taniyama–Shimura conjecture, made by two Japanese mathematicians in the 1950s, and the Kolyvagin–Flach method, developed by Victor Kolyvagin (Russian) and Matthias Flach (German) in the 1980s. Note again the propensity of mathematicians to write the names of their predecessors into the historical record. Note too the complex web of historical interactions behind a single theorem.

Generally speaking, the further back one goes, the more difficult it is to trace the ground between the stepping stones, not least because much of the evidence has long since been washed away.

But without the attempt, there is no history, only the series of anecdotes on which much of the popular history of mathematics is still too often based.

Elite history

Although we know almost nothing about the lives of Euclid or Diophantus, there are just a few things we can say for certain: that both were well educated and could write fluently in Greek, the intellectual language of the eastern Mediterranean; that both had access to earlier writings on mathematics; that both were able to understand, order, and extend some of the cutting-edge mathematics of their day; and that the mathematics they wrote about had no practical value but was a purely intellectual pursuit. The number of men engaged in such mathematics can never have been great, even in a city like Alexandria. Indeed, it has been estimated that at any one time there were no more than a handful of them anywhere in the Greek-speaking world. In other words, both Euclid and Diophantus belonged to tiny mathematical elites.

A moment's reflection is enough to show how much more mathematics must have been going on than the mathematics they wrote about. Greek society, like every other, had its shopkeepers and housekeepers, farmers and builders, and many others who routinely measured and calculated. We know almost nothing about their methods because such people would have learned and taught mostly by example and word of mouth. Nor were they organized into schools or guilds, though we do know of one named group, the *harpēdonaptai*, or rope-stretchers. By its very nature, their mathematics left few traces. Collections of tokens, or marks scratched in wood, stone, or sand, would have been discarded as soon as they were no longer useful, and were certainly not going to be stored in libraries. In any case, these activities were carried out by people of relatively low social status, and were of little or no interest to the intellectuals of the academies.

When mathematical historians speak of 'Greek mathematics', as they frequently do, they are almost always speaking of the sophisticated written texts that have come down to us from Euclid, Archimedes, Diophantus, and others, not of the common or garden mathematics of the *hoi polloi*. Recently this has begun to change. Historians have started to acknowledge that elite Greek mathematics had its roots in the practical and everyday mathematics of the eastern Mediterranean, even if later writers distanced themselves from those roots by developing a more formal and 'useless' kind of mathematics.

There is something else to be wary of in the catch-all phrase 'Greek mathematics'. Diophantus lived in Alexandria in Egypt; Archimedes lived in Syracuse, on the island of Sicily; Apollonius, another of the great 'Greek' mathematicians, lived in Perga, in the region that is now Turkey; in other words, although all wrote in Greek, none of them came from the area that we now know as Greece. Indeed, Diophantus, for all we know, could have been African born and bred. Nevertheless, 'Greek mathematics', so highly revered by Renaissance Europeans, has come to be thought of as essentially 'European'. The absurdity of incorporating Alexandria into Europe becomes all the more apparent when we think of the exclusion of Spain, at the opposite end of the continent. Spain came under Islamic rule early in the 8th century and consequently enjoyed the rich culture and learning of the Islamic world. Yet one frequently reads that Arabic numerals were introduced to Europe by Fibonacci, writing in Pisa in Italy in the early 13th century, as though their use in Spain for two centuries before that counted for nothing, and as though Spain were somehow not a proper part of Europe. Those promoting the cause of elite mathematics have naturally tended to assimilate into their histories whatever would give their subject authority and respectability, regardless of other inconvenient facts.

Wherever mathematics is practised, we are likely to find a few advanced and highly respected practitioners but many more whose names will never enter any history book. If we re-examine the situation in Fermat's day, we will find it hardly any different. During his lifetime, France was exceptionally rich in elite mathematical activity: one can think of as many as three or four Parisians who could have kept up with Fermat. At a generous estimate, there were perhaps as many again in the Netherlands and Italy together, and even one or two in England, but no more than that. Yet mathematical activity lower down the social scale was more widespread than one might expect. Recent electronic searches of digitized material have shown that as many as a quarter of the books published in England in the 16th and 17th centuries mentioned mathematics in one way or another, if only in passing. Further, there was a steady increase in books aimed at tradesmen or craftsmen who wanted to acquire basic mathematical skills.

Before ending this chapter, let us look in a little more detail at one of them: there is after all no better way of exploring the history of mathematics than to delve into the original sources. Robert Recorde's *The Pathway to Knowledg* was published in England in 1551, about 50 years before Fermat was born. For much of his life, Recorde practised as a physician. In 1549 he was appointed controller of the Bristol mint, and two years later surveyor of silver mines in Ireland. Unfortunately he made political enemies in this period and ended up in the King's Bench prison in London, where he died in 1558 at the age of 48. It was also during this time, however, that he published most of the mathematical works for which he is now remembered. Educated at Oxford and Cambridge, Recorde was fluent in Latin and Greek but made the bold choice of writing his mathematical texts in English. In particular, he aimed to make the mathematics of Euclid, one of the most elite of mathematicians, available to the common man. This was no easy task: for one thing, most English workmen, though they may have been adept enough with plumb lines and rulers, had never heard of a formal subject called 'geometry'; for another, there were simply no words in English for technicalities like 'parallelogram' or 'sector'. Recorde addressed both problems with imagination and skill.

In a lengthy preface he described the classes of men for whom geometry was 'much necessary', from those of humblest social status upwards. At the bottom were the 'unlearned sort' who worked the land. Even these men, Recorde argued, had an instinctive grasp of geometry, otherwise their ditches would collapse and their haystacks would topple. Moving upwards to tradesmen, Recorde supplied a long list, in verse, of those to whom geometry was indispensable: merchants, navigators, carpenters, carvers, joiners, masons, painters, tailors, shoemakers, weavers, and more, concluding

> That never was arte so wonderfull witty
> So needefull to man, as is good Geometry.

Recorde also deemed geometry to be indispensable in the professions of medicine, divinity, and law, though his arguments became rather more artificial and less convincing as he climbed the social scale.

Recorde's empathy with the common man is clearest when he comes to the geometry itself: his exposition is a model of good pedagogy, expressed in plain language with lots of examples and helpful diagrams. Quite early on he teaches Euclid's ruler and compass construction of a right angle. In case this should prove too difficult, however, he has an alternative suggestion: take a line and mark off three, four, and five units respectively, and then use those lengths to create a triangle. The angle between the short sides will be a right angle. This is no classical Euclidean construction: it is a method for practical men, for rope-stretchers.

For the 21st century, we could make a far longer list than Recorde could of those who use mathematics in their everyday life, in

1. 'Colourwash' by Tatjana Tekkel Peppé who, by her own account, is no good at mathematics

school, the home, or the workplace. I am thinking of my mother Irene, who at the age of 89 trusted neither banks nor computers, but tallied every penny of her household expenditure in carefully ruled notebooks; or of my friend Tatjana, who repeatedly tells me she was no good at mathematics at school but who creates intricately designed quilts (see Figure 1). She can certainly handle right-angled triangles. Indeed, her instinct for tessellation and proportion qualify her, perhaps, as a modern-day representative of the *harpēdonaptai*.

Elite history does not have any space for Irene or Tatjana: women, in particular, have to rise at least to the level of Sophie Germain before they are taken seriously. Yet without people who do and teach mathematics at every level, the elite could not flourish. Behind the outposts occupied by Wiles, Fermat, or Diophantus, there stretch vast hinterlands of mathematical activity that have been all too little explored in general histories of the subject. Part of the purpose of this book is to redress the balance and to reclaim mathematics for the man, woman, and child in the street, to revisit the history of mathematics from some new perspectives.

Chapter 2
What is mathematics and who is a mathematician?

In the previous chapter, I assumed that readers would take 'mathematics' to be more or less the subject they studied under that name at school, and 'mathematicians' to be those people who continue with it into adult life. History, however, requires us to think about both terms more carefully. So does experience: when as a school teacher I found myself in a single morning delivering lessons on percentages, circle theorems, and differential calculus, I was forced to ask myself how this unlikely collection of topics had come together under the single heading of 'mathematics'. Most people would probably agree with the rather general statement that mathematics is based on properties of space and number, but how then would they regard the popular puzzle Sudoku? Is it a mathematical pursuit or not? I have heard expert mathematicians argue that it is or that it isn't, with equal vehemence either way.

Let us go back to a beginning. The Greek word *mathemata* simply meant 'what has been learned', sometimes in a general way, at other times connected more specifically to astronomy, arithmetic, or music. The Greek word entered into the etymology of the modern word 'mathematics' and its cognates in other European languages (*mathématiques*, *Mathematik*, *matematica*, or, in US English, *math*). The meanings of the word 'mathematics', however, slipped and twisted through many variations over the centuries, as we shall see shortly. And that is looking at the matter only from a European perspective. If we go back one or two thousand years, before European culture became dominant, can we find words equivalent to our 'mathematics' in Chinese, Tamil, Mayan, or Arabic? If so, what writings and activities did those words cover? To investigate that question thoroughly would be a lifetime's work for an army of scholars, but here, as elsewhere in this book, some case studies will serve to illustrate the questions that need to be asked and the kind of answers that can arise.

Tracing some meanings of *suàn*

From histories composed by Chinese government officials for the period from shortly before 200 BC to AD 200 (the Qín and Hàn periods) it is possible to discover the names of just over 20 people who were said to have been skilled in some aspect of *suàn* 算. As a noun, *suàn* can mean a set of short rods, made of wood, metal, or ivory, which are manipulated on a flat surface to record the numbers in a calculation; it can also mean the act of using the rods. Here then is evidence of mathematical activity, but we still do not know very much unless we can discover what kind of calculations were carried out.

For many of the practitioners named in the official records, it appears that *suàn* was closely associated with the astronomical or calendrical systems known as *lì* 曆. All pre-modern societies used the positions of the Sun, Moon, and planets to determine appropriate times and dates for religious rituals or the planting of crops, so those who could make correct predictions from astronomical data were indispensable to rulers and governments. There are thus repeated associations of *suàn* with *lì* in the histories of early imperial China. The same records also show, however, that *suàn* was relevant to more earthly matters, the calculation of profit, and the distribution of resources.

In the early 1980s, a new source was discovered for the period around 200 BC, one that sheds further light on the use of *suàn* at that time. The text known as the *Suàn shù shū* 算 數 書 is inscribed on 190 bamboo strips, each about 30cm long, which were originally joined side by side with knotted string so that they could be rolled together like a mat. The final word *shū* means 'writings' or sometimes 'book'. The middle word, *shù*, may be broadly interpreted as 'number'. What is most relevant for our purposes, however, is the meaning of the combination *suàn shù*. The writings on *suàn shù* contain around 70 problems with instructions for solving them. These include: multiplying whole numbers and fractions; sharing out profit according to the amounts put in by different contributors; allowing for waste in the production of commodities; calculating total cost from the price of a given amount; calculations of tax; finding amounts of ingredients in a mixture; converting an amount of raw material to a number of finished products; checking times taken for a journey; calculation of volumes and areas; conversion of units.

Thus for the most part the problems of the *Suàn shù shū* are based on everyday activities and transactions. It is written in a very direct style: for each problem the writer poses 'question', 'result', and 'method'. Here as examples are two 'customs-post problems' from the second chapter:

A fox, a wild-cat and a dog go through a customs-post; they are taxed 111 coins. The dog says to the wild-cat, and the wild-cat says to the fox 'Your skin is worth twice mine; you should pay twice as much tax!' Question: how much is paid out in each case: Result: the dog pays out 15 coins and $\frac{6}{7}$ coins; the wild-cat pays out 31 coins and 5 parts; the fox pays out 63 coins and 3 parts. Method: let them be double one another, and combine them [into] 7 to make the divisor; multiply each by the tax to make the dividends; obtain one for [each time] the dividend accommodates the divisor.

And perhaps more realistically:

> A man is carrying hulled grain – we do not know how much – as he
> passes through three customs posts. [Each] post takes a duty of 1 in
> 3. After leaving he has one *dǒu* of hulled grain left. Question: when
> he started going, how much hulled grain did he bring? Result: The
> hulled grain he brought was 3 *dǒu* 3 *shēng* and $\frac{3}{4}$. Method: Set out
> one, and thrice double it to make the divisor. Again set out one *dǒu*
> of hulled grain and 3-fold it. Again three-fold it and [multiply by]
> the number of passes to make the dividend for it.

The answers are correct but the descriptions of the 'method' are
not very enlightening, and it is likely that they were meant to be
supplemented by oral explanation. The instructions are given only
for the particular numbers in the stated question, but a trained
reader would be able to adapt them to any similar problem, so in
that sense they provide a general technique. There is no
expectation in the text, however, that the reader should
understand the reasoning behind the method, only that he should
be able to apply it.

Similar problems and others appear in a later text, the *Jiǔ zhāng
suàn shù* 九章算術, writings on *suàn shù* in nine sections,
commonly known in English as the 'Nine Chapters'. The official
histories show that the text was in use by the beginning of the 2nd
century AD. As with Euclid's *Elements* from some three or four
centuries earlier, however, we have no information about the
author or composition of the 'Nine Chapters', nor the original text.
The only version that has come down to us is the one given by Liú
Huī 劉徽 in AD 263. Until the transcription and publication of the
contents of the *Suàn shù shū* in 2000, the 'Nine Chapters' was the
earliest extensive text devoted to *suàn*. The discovery of the *Suàn
shù shū* therefore not only makes possible important
textual comparisons, but also offers historians a much deeper
knowledge of the uses of *suàn* in the early years of imperial
China.

It is clear even from this very short account that the word *suàn* was not associated with any overarching subject that we can capture with the single word 'mathematics'. Instead it denotes techniques and skills that could be put to use in a range of contexts, from applications to *lì*, the astronomical reckonings required at court, to the more mundane *suàn shù*. Turning now to the Latin West, can we find a similar range of practices associated with the word 'mathematics'?

Tracing some meanings of 'mathematics'

Around AD 100 the Roman writer Nicomachus listed four disciplines concerned with multitude and magnitude: arithmetic, music, geometry, and astronomy. For Nicomachus, arithmetic, the study of multitudes (or numbers), and geometry, the study of magnitudes, were the most fundamental; music was the science of multitudes in relation to one another, while astronomy dealt with magnitudes in motion. Four centuries later, the philosopher Boethius described these disciplines collectively as the *quadrivium*. Together with the *trivium* of grammar, logic, and rhetoric, they made up the seven liberal arts of the medieval academic curriculum. Boethius himself wrote treatises on arithmetic and music that were studied in European universities throughout the medieval period. Some writings on geometry were also ascribed to him, but their true authorship is uncertain: Boethius, like Pythagoras, became something of a mythical figure to whom later work could usefully be attributed.

Arithmetic and geometry remain at the heart of mathematics (they are the activities, we may recall, practised by Irene and Tatjana) but astronomy and music have now gone their separate ways. The break came in the 17th century when it became increasingly difficult to reconcile mathematical theory with musical practice, and when astronomy struggled to free itself of its long associations with astrology to become a respectable subject in its own right.

In any case, by the time of the Renaissance the four-fold division of Nicomachus was too constrained to accommodate the many new kinds of mathematical activity that were beginning to emerge in response to the rapid growth of wealth, trade, and travel. John Dee, in a preface to the first English translation of Euclid's *Elements*, in 1570, set out a 'groundplat', or plan, of the mathematical arts and sciences (see Figure 2). Arithmetic and geometry remain the key components, but by now geometry, which answers the questions 'How farre?', 'How high or deepe?', 'How broad?', has given rise to 'geographie', 'chorographie', 'hydrographie', and something called 'stratarithmetrie'. Further, there is a long list of subjects regarded as 'derivatives' of both arithmetic and geometry, including 'astronomie' and 'musike' among many others. The modern reader will have some idea of what was meant by 'perspective', 'cosmographie', 'astrologie', 'statike', 'architecture', and 'navigation' but will probably be as bemused as contemporary readers might have been by 'anthropographie', 'pneumatithmie', 'archemastrie', and several other uncommon branches of learning. Indeed, the obscurity of the subject matter and the neat divisions under subheadings and sub-subheadings suggest that Dee's systematization, like the much simpler scheme of Nicomachus or Boethius, was a philosophical exercise rather than a genuine classification of existing practices.

How then are we to find out more precisely what mathematical activity in western Europe consisted of during the centuries between AD 500 and 1500? Can we carry out the same kind of study for 'mathematics' as we did for *suàn*, discovering the meanings of the word by examining the contexts in which it was used? There are many more surviving texts from western Europe over this period than there are for early imperial China, so a full survey is impossible, but as a first approach we will examine a mathematical history compiled by the Dutch scholar Johann Gerard Vossius, his *De scientiis mathematicis*, published in Amsterdam in 1649, in particular as it relates to British writers.

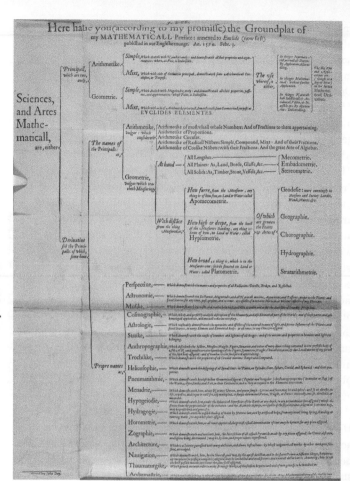

2. John Dee's 'groundplat' from his preface to Euclid's *Elements*, 1570

It may seem strange to turn to a Dutch scholar for information on British intellectual history, but much of Vossius's account of British authors was based on earlier work by the English antiquary John Leland. In 1533, shortly before the dissolution of the monasteries, Leland was commissioned by Henry VIII to search the libraries and colleges of the realm and to list their collections. Over the next two or three years, he listed the holdings of some 140 religious foundations. The subsequent dispersal and loss of books grieved him greatly: in 1536 he complained to Thomas Cromwell that 'the Germanes perceiving our desidiousness [indolence] and negligence, do send dayly young Scholars hither, that spoileth them, and cutteth them out of Libraries'. Leland provided the last and most comprehensive record of what the libraries had contained. He intended to compile a dictionary of British writers, containing some 600 entries, but sadly became insane before it was quite completed. His invaluable work was recognized by other historians, however, and a great number of later writers, including Vossius, drew directly or indirectly on his findings.

The earliest English writer mentioned by Vossius was Bede, writing about AD 730, who was listed under both 'astronomy' and 'arithmetic'. Bede, who spent most of his life in the monastery at Jarrow in the north-east of England, is well known as a biblical commentator and church historian, but few would now count him an astronomer. The writings ascribed to him, however, are described as being on the Moon and its cycles, the date of Easter, the planets and zodiac, the use of the astrolabe, and calculation of the vernal equinox. Some of these writings may have been mistakenly ascribed to Bede by later commentators, but he was certainly much concerned with the date of Easter, which was as crucial to Christians as the correct timing of the winter solstice had been to the early Chinese emperors. It was not an easy calculation either: Easter had to fall on the first Sunday following a full Moon after the spring equinox, and so correct calculation of the date required an understanding both of lunar and solar cycles, which are not naturally correlated. The presence of two Christian traditions in northern England, Irish and Roman, had led to conflicting dates, a situation that had eventually been resolved at the Synod of Whitby in 664. Bede may not have carried out the necessary calculations himself, but he knew what was at stake.

Calculations concerned with ecclesiastical time-keeping eventually became known by the name of *computus*, and remained essential throughout the medieval period.

After Bede and his follower Alcuin, no further English names appear in Vossius's account for more than four centuries, until we meet Adelard of Bath around 1130. Adelard, who appears to have travelled in France, Sicily, and Syria, was one of the earliest translators of parts of Euclid's *Elements* from Arabic to Latin, and was also said to have written on the astrolabe.

Only for the 13th and 14th centuries do the names (and their supposed dates) begin to appear with increasing frequency, all of them under 'astronomy' or 'astrology': John Sacrobosco (1230), whose writings on the Earth and its place in the universe remained a key part of the university curriculum for four centuries; Roger Bacon (1255), described as an astrologer; Walter Oddington (1280), said to have written on the motion of the planets; Robert Holcot (1340) of Northampton, said to have written on the motion of the stars; John Eastwood (1347), astrologer; Nicholas Lynne (1355), astrologer; John Killingworth (1360), astronomer; Simon Bredon (1386), said to have written on medicine, astrology, and astronomy; John Summer (1390), astrologer, and so on. Then in the 15th century the names begin to fade out again. Clearly the 14th century was a high point of astronomical and astrological studies, one contributory factor perhaps being the terrible shock of the Black Death in 1348. Many of those mentioned belonged to religious orders, Franciscan, Dominican, or Carmelite. Many were also linked to Oxford, particularly to Merton College, and some of their writings are safeguarded to this day in Oxford libraries. All of them crossed and re-crossed the fluid boundaries between astronomy and astrology.

In contrast to this galaxy of astronomers, no English writers appear in any of Vossius's chapters on music, optics, geodesics, cosmography, chronology, or mechanics, and only Gervase of

Tilbury and Roger Bacon are mentioned under geography, as map-makers. Thus, looking back from a 16th-century standpoint to the mathematical writings of medieval England, the dominant themes are *computus* and astrology.

For other regions of Europe, however, the picture would have been different. In Italy, for instance, situated in the heart of the western Mediterranean, trade was more extensive and more complex than in northern Europe, and the 13th century saw the establishment of *abacus* schools to train boys in commercial arithmetic and even a little rudimentary algebra (solving some basic equations). The seminal text was the *Liber abaci* (*Book of Abacus*) of Leonardo of Pisa, later known also as Fibonacci. The *Liber abaci* contains hundreds of commercial problems. Here are two of them:

Four men made a company in which the first man put $\frac{1}{3}$ of a whole, another put $\frac{1}{4}$, a third put $\frac{1}{5}$, and a fourth truly put $\frac{1}{6}$, and they had together a profit of 60 soldi; it is sought how much each held of it. The problem truly is the same as was said about four men who buy a pig for 60 soldi of which the first wishes to have one third of the pig, the second one fourth, the third one fifth, and the fourth one sixth....

Leonardo himself has pointed to two versions of this question; it is also mathematically equivalent to the fox, dog, and wild-cat problem from the *Suàn shù shū*. The next problem reflects the concerns of contemporary Italy and is typical of hundreds of questions on conversion of currencies or materials. At the same time, it shows that some ten centuries after Diophantus, arithmetic of another kind was still thriving in Alexandria.

Also 11 Genoese rolls [of cloth] are worth 17 carats in Alexandria; how much are 9 Florentine rolls worth? Because the 11 rolls and the 9 rolls are not the same units of weight, you make Florentine rolls of the 11 Genoese rolls, or you make Genoese rolls of the 9 Florentine rolls so that both will be either Florentine rolls or Genoese rolls; but because you can easily make Florentine rolls, each Genoese roll is $\frac{1}{6}2$ Florentine rolls, you will multiply the Genoese rolls by $\frac{1}{6}2$ to make $\frac{5}{6}23$ Florentine rolls...

For all their learning, Vossius and his sources in northern Europe had never seen the *Liber abaci*; Vossius knew of it only by hearsay and got its date wrong by two centuries. Mathematical activity could be very localized.

It was also time-related. For the medieval period, most of the headings later invented by Dee and Vossius would have been largely redundant, at least for England. By the end of the 16th century, as Britain too entered into the wider world, that was no longer the case. Thomas Harriot, working around 1600, left writings on optics, ballistics, alchemy, algebra, geometry, navigation, and astronomy. Meanwhile, his contemporary Simon Stevin in the Netherlands published on a similar range of subjects, but with navigation replaced by the more pertinent problems (to him) of locks and sluices. *Computus* and astrology had given way to the mathematical activities of a new world order.

What is mathematics?

What then has mathematics been historically, if indeed there has ever been such an entity? It should be clear by now that mathematical activity has taken many forms, only loosely connected by the fact that they require some kind of measurement or calculation. A more precise answer must be heavily dependent on time and place. There are a few common threads: all organized societies need to regulate trade and time-keeping, which were very roughly speaking the aims of *suàn shù* and *suàn lì*, respectively, in early imperial China, or of *abacus* and *computus* in 13th-century Europe. The practitioners of these various techniques, however, were likely to have been of very different social status. *Suàn shù* and *abacus* teachings were intended for merchants or officials, whereas *suàn lì* or *computus* were the province of high-ranking specialists in China, and of monks and scholars in medieval Europe. A separation of status and respect between those sufficiently educated to engage in 'higher' mathematics, which usually requires a certain level of abstract thinking, and the tradesmen or craftsmen who work with 'common' or 'vulgar' mathematics, has recurred in different contexts over many centuries.

As societies become more complex so too do their mathematical requirements. The long list of headings proposed by Dee, even if some were redundant, indicates a wide range of activities in which mathematical expertise was invoked. These subjects were collectively known as 'mixed mathematics', suggesting that mathematics was an integral part of each of them (not quite the same as the later concept of 'applied mathematics', where mathematics is used to analyse subjects outside itself).

There is no reason to suppose that the lessons learned from early imperial China and medieval Europe do not extend to other societies too: that there is no single body of knowledge that we can conveniently call 'mathematics' but that we can identify many mathematical disciplines and activities. And which particular ones are considered most relevant or prestigious has always been a matter of time and place.

Who is a mathematician?

Now that we have begun to identify the range of activities that have constituted mathematics, can we say who does or does not count as a mathematician? All four of Pythagoras, Diophantus, Fermat, and Wiles are commonly described as mathematicians, and the first three, being dead, have made it into a standard reference work, the *Biographical Dictionary of Mathematicians*. None of them, however, would have recognized the label they have been given. We have no idea how, if at all, Pythagoras would have described himself. Diophantus would probably have thought of himself as an arithmetician, not as a practitioner of everyday arithmetic of the *suàn shù* or *abacus* kind, but of the 'higher arithmetic' that investigates some of the more obscure and difficult properties of the natural numbers. Fermat, on the other hand, would have called himself a *géomètre*, geometry by then being the most authoritative and respectable branch of the quadrivium. This remained a standard description of an academic mathematician in France well into the 19th century. Of the four, only Wiles, I suggest, would unreservedly call himself a mathematician.

Today the discipline of mathematics is highly respected, even revered, but from what has been said already in this chapter it can easily be seen why this has not always been the case. John of Salisbury in the 12th century claimed that the practice of *mathematica*, the foretelling of the future from the positions of the stars and planets, arose from a fateful familiarity between men and demons, and along with chiromancy (palm-reading) and augury (interpreting the flight of birds) was a source of evil. In 1570, Girolamo Cardano, medical practitioner and author of one of the leading algebra texts of the Renaissance, was imprisoned for casting a horoscope of Christ; Thomas Harriot, arrested in 1605 on charges of association with perpetrators of the Gunpowder Plot, was questioned not so much about the plot itself but about the fact that he had a horoscope of James I pinned to his wall; and late in the 17th century John Aubrey wrote of the country clergyman and mathematics teacher William Oughtred that 'The country people did believe that he could conjure'. In pre-modern Europe, the practice of 'mathematics' was not without its dangers, to the practitioner as much as to his supposed subjects.

In fact, the word 'mathematician' began to be used regularly in English mathematical writings only from 1570 onwards. At first, it was used mainly for foreign authors, but later in two curiously unrelated contexts, for gunners or astrologers. After the Restoration in 1660 it came to be used more generally for writers on arithmetic or geometry but also still for astrologers; at the same time, the predictions of 'the mathematicks' became a regular subject of satire and ridicule. The longstanding and persistent association of mathematics with astrology helps to explain why academics preferred to avoid the term. When Henry Savile founded two mathematical chairs in Oxford in 1619, they were in Geometry and Astronomy, respectively, with strict directions that the latter should not include judicial astrology. To this day Cambridge hosts a Lucasian Professor of Mathematics, but Oxford's equivalent is the Savilian Professor of Geometry. And unless it should be thought that the association of mathematics with prediction and influence was only a European phenomenon, it is worth bearing in mind that the modern Chinese term for mathematics, *shù xué* 數學, has traditionally meant the study of numbers in the context of divination.

In short, 'mathematicians', as we now understand the term, are a modern European invention. In the long history of mathematical activity, they have existed for little more than a blink of the eye, and if we are to appreciate mathematical history properly it is crucial not to project their image back onto the past. For that reason, historians prefer to use more precise descriptions like 'scribe', 'cosmographer', or 'algebraist', or more general terms like 'mathematical practitioner'. One thing is certain: the history of mathematics is not the history of mathematicians.

Chapter 3
How are mathematical ideas disseminated?

The previous chapter carried out some broad surveys of mathematical activity at different times and places. This is one way of studying the history of mathematics: determining what people actually did. The historian always wants to ask further questions, however: not just what people knew but how they communicated it to one another and to those who lived after them. How are mathematical ideas passed from one person to another, from one culture to another, or from one generation to another? (Recall the questions first raised in Chapter 1: how did Fermat know about Diophantus, or Wiles about Fermat?)

An extension of these questions is to ask how historians themselves can know about the mathematics of the past: what sources do we have, how have they come down to us, how reliable are they, and how can we learn to read them? This chapter will examine the way mathematical ideas have sometimes traversed long distances of time and space, but also how, often, they have not.

Fragility, scarcity, and obscurity

Those who have comfortably assumed that mathematics began with Pythagoras may now suffer a slight sense of vertigo on discovering that sophisticated mathematics was already being practised more than a thousand years earlier in Egypt and in the region that is modern Iraq. The Egyptian and Babylonian civilizations of the 2nd and 1st millennia BC existed in relative proximity to each other, but we know very much more about the mathematics of the latter than the former for the very simple reason that the clay tablets used as writing material along the Tigris and Euphrates were robust and durable whereas the papyri of the Nile region were not. Thousands of tablets have been excavated from Iraq, many with mathematical content, and thousands more probably remain buried if they have not been crushed by tank treads or looted in the chaotic aftermath of recent wars. For Egypt, on the other hand, the number of surviving mathematical texts and fragments can be counted on the fingers of three hands, and those are scattered across a thousand years of history. The equivalent for Britain would be a few texts from around the time of the Norman conquest and a few more from the 19th century. Clearly the surviving Egyptian texts provide only the most meagre insight but at the same time leave ample room for speculation and fantasy about Egyptian mathematical activity.

For India, south-east Asia, and South America, the situation has been much the same as for Egypt: the climate has rapidly destroyed natural materials like wood, skin, or bone, so that historians have had to do the best they can with very few texts, poorly preserved. Clearly the paucity of material distorts our picture of the past. We must ask whether what survives is typical of what has been lost, knowing that a single new discovery (like the *Suàn shù shū* in China) could radically alter our perceptions of an entire mathematical culture. At the same time, the lack of texts has perhaps had some benefits in that it has forced historians to broaden their search for sources. Administrative records, for instance, can reveal the counting and measuring that were carried out in everyday life. Archaeological evidence has improved our knowledge of how buildings were planned and constructed and what calculations must therefore have gone into them (for we have no direct evidence of any calculations that went into the building

33

of Stonehenge or the Pyramids). Sources as varied as pictures, stories, or poems may also include hints of contemporary mathematical knowledge.

Many ancient texts were written in scripts and languages that are now extinct, and the process of translating them is fraught with difficulties. The number of scholars with the requisite language skills who are also brave enough to engage with mathematical material remains very small indeed, and their task is exceedingly delicate. Any translation from one language to another risks destroying something of the essence of the original, but mathematical translation introduces a further difficulty: how to make the technical concepts of another culture comprehensible to a modern audience. What can the ordinary reader make, for example, of the following passage from the Indian *Brāhmasphuṭasiddhānta* from AD 628:

> The height of a mountain multiplied by a given multiplier is the distance to a city; it is not erased. When it is divided by the multiplier increased by two it is the leap of one of the two who make the same journey.

To understand this problem, the reader needs to know that one traveller descends a mountain and walks along the plain to a city, while the other magically leaps from the mountain top to a greater vertical height and flies along the hypotenuse, but in doing so covers just the same distance. For a student at the time, this problem may have been one of a standard type (another version of it has monkeys leaping up trees) and was probably elucidated through oral explanation, but for a 21st-century reader with no knowledge of Sanskrit or 7th-century Indian mathematical conventions, it is at first sight simply baffling.

Thus a literal translation of a raw text is not likely to convey very much to a non-specialist. An age-old way of getting round this problem is for translators (or copyists) to add annotations or explanatory diagrams: all important mathematical texts have accrued layers of commentary in this way. Another method is to translate the text into modern mathematical notation. The reader who wishes to try this for the problem of the two mountain travellers will probably find that it makes it much clearer. The use of modern algebraic notation can be helpful as a preliminary way in to understanding the mathematics of the past but should never be mistaken for what the original writer was 'really' trying to do, or what he would have done with the advantage of a good modern education. At best, such modernization obscures the original method and at worst can lead to serious misunderstandings.

The surviving Egyptian texts of the 2nd millennium BC, for example, are written in hieratic, a cursive script that replaced hieroglyphics in everyday use from about 2000 BC onwards. They were translated into English or German in the early 20th century and for many years those translations remained standard. Unfortunately, however, the contents were translated not only into modern languages but also into modern mathematics. It is often stated, for instance, that the Egyptians used a value of 3.16 for the number we now denote by π, the multiplying factor that gives the area of a circle from its radius-squared (as a modern formula, we may write $A = \pi r^2$). When we examine the texts on which this claim is based, we find that they do not expect the reader to multiply the radius-squared by any number at all. Instead they instruct him to find the area by reducing the diameter by $\frac{1}{9}$ and then squaring it. A bit of pencil and paper calculation shows that this gives the area of the circle as $\frac{256}{81}$ times the radius-squared, hence the magic value of $\frac{256}{81} = 3.16\ldots$ But 'reducing and squaring' is not the same as 'squaring and multiplying', even if it gives very nearly the same answer: the process is quite different, and processes are precisely what historians need to be concerned with if they are to understand the mathematical thinking of earlier cultures.

The story of translation of Babylonian texts has been similar. Here the languages are Sumerian, unrelated to any surviving language, and Akkadian, a precursor of Arabic and Hebrew; and the writing is cuneiform, imprinted into wet clay with a sharpened reed. A large number of mathematical texts were translated and published during the 1930s by Otto Neugebauer and François Thureau-Dangin, and thereafter for many years the job was thought to be more or less done. These early translations, however, all too often turned Mesopotamian calculation techniques into their modern algebraic equivalents, obscuring the true nature of what the original scribe was actually thinking and doing, while at the same time making the calculations look rather primitive. Only since the 1990s have many of the tablets been translated afresh, with much greater care for the original language. Words that mean literally 'break in half' or 'append', for instance, convey physical actions that are quite lost in the abstract translations 'divide by 2' or 'add', and give us a much better insight into the way problems were understood or taught.

Reading and translating texts is only one part of the work of historians of ancient mathematics, albeit an important one. The other is to interpret them within their own context. Sometimes this is simply impossible: many Middle Eastern texts excavated or rediscovered in the 19th century, including almost all the extant Egyptian hieratic texts and hundreds of Old Babylonian cuneiform tablets, changed hands in the antiquities market carrying no known provenance. Unfortunately, many looted or stolen objects are still bought and sold this way today.

The fragility and scarcity of mathematical texts improves only a little as we move forward from the ancient world to the medieval period. Even documents deliberately preserved in libraries are not always secure. There are varying accounts, now impossible to confirm, of the destruction of the library at Alexandria in times of conflict, and certainly it would have been as vulnerable to fire as any pre-modern library housing books or manuscripts. Readers at

Oxford's Bodleian Library are still required to swear an oath promising 'not to bring into the Library, or kindle therein, any fire or flame, and not to smoke in the Library', a reminder of the days when such activities could prove as lethal to books as to people.

We have already seen the efforts of John Leland to record the contents of monastic libraries, but he could not preserve more than a fraction of the collections themselves when those libraries were eventually destroyed and their holdings dispersed. There were other dangers too: Merton College in Oxford threw out a great many manuscript books during the 16th century as it modernized to printed texts, and though some were rescued by alert collectors, there must have been many that were not. And John Wallis in 1685, like Leland more than a century earlier, complained bitterly about the theft of valuable material: two 12th-century prefaces, he wrote, had 'lately (by some unknown hand) been cut out, and carried away' from a manuscript in Corpus Christi College. He hoped that 'Who ever hath them, would do a kindness (by some way or other) to restore them', but he hoped in vain: the prefaces are still missing.

Private collections of papers were also vulnerable: John Pell, worrying in 1644 about the mathematical papers of his recently deceased friend Walter Warner, wrote:

> I am not a little afraid that all Mr Warner's papers, and no small share of my labour therein, are seazed upon, and most unmathematically divided between the sequestrators and creditors, who will, no doubt, determine once in their lives to become figure-casters, and so vote them all to be throwen into the fire.

Printed books are just as susceptible as manuscripts to fire, flood, insects, and human carelessness, but because more copies are produced more are likely to survive. Those that come down to us, however, are unlikely to be typical of what once existed. An expensive volume from a gentleman's library is more likely to last than a tradesman's well-thumbed ready-reckoner, but probably tells us less about what was actually read and used.

Constructing a true understanding of the past often feels like trying to put together a jigsaw puzzle in which most of the pieces are missing and there is no picture on the box. Nevertheless, remarkably, we do have mathematical texts that have survived for centuries, even millennia. For the most part, their contents are of purely historical interest: no-one now calculates with Egyptian fractions, except as a school exercise, and the only vestiges of the Babylonian sexagesimal system are in our otherwise curious division of an hour into 60 minutes and a circle into 360 degrees. Other texts, however, have remained very much alive through continued use and translation, and occasionally it is even possible to trace an almost continuous line of descent from past to present. The outstanding example must be Euclid's *Elements*, which has already been mentioned more than once, and without which no history of mathematics can be complete. A study of what is sometimes called the 'transmission history' of the *Elements* tells us much about how mathematical ideas from the past may be preserved, amended, and passed on.

Preservation through time

The remarks made above on the fragility of Egyptian sources apply just as much to texts from the ancient Greek-speaking world, which were also written on papyrus. We assume, from contemporary references to some of his other works, that Euclid wrote around 250 BC. Yet the earliest surviving text of the *Elements* is from AD 888. That represents over a thousand years of copying and re-copying, with all the scope that entails for error, change, and 'improvement'. How can we know that the text we have now is in any way faithful to the original? The answer is that we can not. In the case of the *Elements*, we have extensive commentaries from later Greek writers, Pappus (AD 320), Theon

(AD 380), and Proclus (AD 450), which tell us how the text appeared in the 4th or 5th century BC. These men lived much closer to Euclid's time than we do, but still several centuries after the *Elements* was first written. Otherwise, the only way historians can approach the original is to construct a 'family tree' of surviving manuscripts, by observing, for example, where mistakes or alterations have been copied from one text to another. In this way they would hope to arrive back at a 'master copy', but it is painstaking work with no guarantee that it will take one back to a true and unique source.

That earliest surviving manuscript of the *Elements*, from AD 888, is written in Greek and was preserved in Byzantium. But as Islam spread into the old Greek-speaking regions of the Mediterranean, the text was also translated into Arabic. One can imagine what difficulties the early Islamic translators might have encountered by comparing their task with that of Robert Recorde several centuries later: it is unlikely that Arabic, the language of nomadic tribes, contained ready-made words for the abstract concepts of Euclidean geometry. Nevertheless, Arabic translators saved many texts from extinction.

Most medieval translations of the *Elements* into Latin were thus made not from Greek, a language that had by then all but died out in western Europe, but from Arabic sources in Spain or Sicily. Adelard of Bath, whom we met in the previous chapter, was one such translator, and there were several others in the 12th century, scholars from northern Europe who travelled south in search of the learning that could be found there. Eventually, as knowledge of Greek was slowly revived, translations were also made directly from Greek sources.

Once printing was established, in the 15th century, Euclid's *Elements* was finally secured for posterity. It was amongst the first mathematical books to be printed, in a beautiful edition of 1482 that continued the traditions of manuscript production: there is

3. The first page of the first printed edition of Euclid's *Elements*, 1482

no title page (because manuscript writers traditionally signed their names at the end of a text, not at the beginning), and it contains delicately painted illuminations (see Figure 3).

During the 16th century, printed editions followed each other rapidly, at first in Latin and Greek but then in several vernacular languages. Robert Recorde included most of the material from the first four books of the *Elements* in *The Pathway to Knowledg* in

1551, and some further and more difficult material from the later books in his last publication, *The Whetstone of Witte*, in 1557. The first full English translation of the *Elements* was published in a lavish edition by Henry Billingsley in 1570: it contains Dee's 'groundplat' and is also the earliest known English text to display the word 'mathematician' on the title page.

Over the next four centuries, there were many further translations and editions as editors adapted to the changing needs of the time. By the mid-20th century, the *Elements* was finally eased out of the school curriculum (though not its contents: schoolchildren still learn to construct triangles and bisect angles). It has not, however, disappeared from the public domain. A modern interactive web version is the most recent innovation in a very long tradition of translating and adapting the *Elements* for each new generation.

The *Elements* has been unique in its reach and longevity, but the story of its preservation is typical of that of many other Greek texts, including the *Arithmetica* of Diophantus, from which Fermat's Last Theorem arose. A similar story about early commentaries, translations into Arabic, later translations into Latin, and eventual print publication from surviving Greek sources can be told for most classical texts. There has been just one exception, the near miraculous rediscovery in the early 20th century of an otherwise lost text by Archimedes, faintly discernible below later writing and paintings on the pages of a Byzantine prayer book. Such finds are exceedingly rare, and serve to remind us yet again of how much mathematics of any culture has also been lost.

Preservation over distance

Despite the fragility of written documents, mathematics has been communicated not just over long periods of time, but sometimes over long distances, and sometimes both. We begin with a mystery.

Here is the beginning of a problem from an Old Babylonian tablet now in the British Museum (BM 13901):

I summed the area and my square-side so that it was 0;45.

Using the technique that was warned against above, let us introduce algebraic notation for just long enough to see what the problem is about. If we let the side of a square be s then its area is s^2. The number 0;45 is a modern transcription that we may interpret as $\frac{45}{60}$ or $\frac{3}{4}$. Thus the statement can be written in modern terms as the equation $s^2 + s = \frac{3}{4}$. The Babylonian technique for finding the length of the square-side involved slicing and rearranging geometrical shapes; for the trained practitioner, this could be reduced to a series of brief instructions, a recipe guaranteed to give the answer.

Now consider this problem from a text on the subject of *Al-jabr wa'l-muqābala* ('Restitution and Balancing') composed by al-Khwārizmī in Baghdad around AD 825.

A square and 21 units are equal to 10 roots.

Here the 'roots' are the square-roots of the given square, and so if we once again use modern notation we see that the problem can be written as $s^2 + 21 = 10s$. In other words, this is closely related to the Old Babylonian problem written down more than two and a half thousand years earlier. Further, al-Khwārizmī gave a very similar recipe for finding the answer. His text was so influential that it gave its name to the subject now known as algebra.

Is it coincidence that the same kind of problem with the same kind of solution reappeared so many centuries later in the same part of the world? There is no evidence at all for continuity down the years as we have for Euclid's *Elements*, certainly not within ancient or Islamic Iraq. We do, however, have evidence of ideas being carried from late Babylonian culture to India, and of mathematics later being transported in the other direction, from India to Baghdad. It is just possible that problems like those discussed here were part of that flow: we cannot say and can only speculate. It is worth rehearsing, however, what we know with more certainty.

From about 500 BC to 330 BC, ancient Iraq and north-west India were distant partners in the Persian empire, after which for a short time the same region came under the rule of Alexander the Great. Evidence for the absorption of Babylonian mathematics into India is circumstantial but fairly clear, especially in astronomical calculations: it can be seen in the Indian use of base 60 in measurements of time and angle, and in similar methods of calculating the length of daylight throughout the year. (In India, as in other early societies, correct time-keeping for ritual and other purposes was essential.) Later, there were translations into Sanskrit of Greek astronomical or astrological texts, so that the Greek 'chord', used in measuring astronomical altitude, became the basis of the Indian 'sine'. The dearth of early Indian texts prevents us from knowing what other knowledge must have passed eastwards, and no doubt in the other direction too: a few astronomical fragments from pre-Islamic Iran, for example, suggest the influence there of Sanskrit texts.

By the end of the 6th century AD (or even much earlier) there had been developed in parts of central India a system of writing numbers using just ten digits together with a system of place value. The importance of this can hardly be overstated. In modern parlance, it means that we can write any number of any size (or smallness) using just the ten symbols 0, 1, 2, 3, 4, 5, 6, 7, 8, 9. 'Place-value' means that '2' and '3' stand for different values in 200,003 and 302 because they are positioned differently. In both cases, the zeros serve as place-holders so that we do not mistake 200,003 for 23 or 302 for 32. Once this has been understood, the same few rules for addition and multiplication can be applied to numbers of any size. Of course, there have historically been many other ways of writing numbers, but all of them require the invention of more and more new symbols as the numbers get larger, and none is convenient for pencil and paper calculation: try adding a pair of numbers written in Roman numerals, xxxiv and xix for instance, without converting them into something more familiar.

The Indian or Hindu numerals, as they came to be called, were already known in parts of Cambodia, Indonesia, and Syria as early as the 7th century: they were highly praised by the Syrian bishop Severus Sebokht, for example. By AD 750 Islam had spread over the area of the old Persian empire (and beyond); and by 773 the Hindu numerals had arrived in Baghdad in astronomical treatises brought to the Caliph al-Manṣūr from India. Around 825, al-Khwārizmī, whom we have already met as a writer on *al-jabr*, wrote a text on the use of Indian numerals. The original is lost, but its contents can be recovered from later Latin translations. It taught first how to write the ten digits, in their Arabic rather than Sanskrit forms, with careful explanation of place-value and the correct use of zero; this was followed by instructions on adding and subtracting, doubling and halving, multiplication and division, some teaching on fractions including the sexagesimal kind, and directions for extracting square-roots. Al-Khwārizmī's text set the pattern for arithmetic texts for centuries: its outline can still be easily discerned in many 17th-century European texts even though the material was by then much expanded. For now, however, let us stay with the Indian numerals themselves or, as they had now become, the Hindu–Arabic numerals, as they continued to spread westward.

By the end of the 10th century, the numerals had been carried to Spain, at the other end of the Islamic world from India, and had acquired the western Arabic form that prefigured modern western numerals, rather than the eastern Arabic form still used in Arabic-speaking countries. And from Spain, they were slowly disseminated northwards into France and England. One of the myths about the numerals is that they were introduced into Christian Europe by a monk called Gerbert, later Pope Sylvester II, who had visited Spain before 970. It is true that Gerbert used the numerals on abacus counters, but on this slender evidence one can hardly give him credit for introducing them to the rest of Europe: we do not know whether he had learned the relevant methods of calculation or whether he merely used the numerals as decorative symbols; nor do we know how widely his abacus was known or used; and besides, there must have been other travellers to Spain who similarly brought back a little knowledge of the numerals to demonstrate to their friends. Knowledge of the numerals probably spread only slowly and in a piecemeal sort of way until their usefulness began to be better recognized.

We do know that astronomical tables from Spain, the Toledan tables, were adapted for Marseilles in 1140 and for London in 1150. The instructions for using the tables were translated from Arabic to Latin but the tables themselves were not: who would want to convert columns of two-digit figures measuring degrees, minutes, and seconds into clumsy Roman numerals? Just as astronomical tables had carried the Indian numerals to Baghdad, so they later brought them to northern Europe: for astronomers, the numerals were not just useful but crucial in order to make sense of other people's observations.

At a more mundane level, knowledge of the numerals and associated methods of calculation must also have spread westwards and northwards through trade. The Crusaders, for instance, would have encountered them from the late 11th century onwards. Unlike astronomical tables, however, records of buying and selling were ephemeral and have long since vanished.

By the 12th century, texts were being written specifically to explain the new numerals and the associated methods of calculation. One of them was Leonardo of Pisa's *Liber abaci* which circulated in Italy but not in northern Europe. In France and England, there were instead to be found Latin texts called 'algorisms', the name being a corruption of their opening words, 'Dixit Algorismi', meaning 'Thus spake al-Khwārizmī'. These texts, like al-Khwārizmī's original treatise, taught how to write the numerals and how to carry out basic arithmetic with them. A particularly charming one, known as the 'Carmen de algorismo', was composed in verse by Alexander de Ville Dieu from northern France. The opening lines are, in translation:

> This present art is called algorismus, in which
> We make use of twice-five Indian figures:
> 0.9.8.7.6.5.4.3.2.1.

Alexander went on to explain how the position of each numeral mattered:

> If you put any of these in the first place,
> It signifies simply itself: if in the second,
> Itself tenfold.

Despite their obvious advantages, the uptake of the numerals was slow, not, as is sometimes suggested, because of their Oriental and non-Christian origins, but because for everyday use the old Roman system together with calculations done on fingers or counting boards served well enough. Besides, not everyone found the new numerals easy to learn: as late as the 14th or 15th century a monk in the Benedictine monastery of Cavenso in Italy numbered some of his chapters from the 30th onwards as XXX, XXX1, 302, 303, 304, ... Eventually, however, the Hindu–Arabic numerals superseded all others, and once they had been taken west to America had almost completed their circumnavigation of the world.

There are other stories that could be told about the way mathematics has been disseminated over long distances. Traditional Chinese mathematics, for instance, was taken up by all China's immediate neighbours, and there were no doubt exchanges with India as well, but none with the West until the Jesuits arrived in the 17th century, bearing the *Elements* of Euclid. Such movements have continued in more recent times: in the 19th century, European mathematics was carried outwards from its heartlands in France and Germany to the peripheries of Europe, the Balkans at one end and Britain at the other, and then to the United States, and eventually to every country in the world. Such dissemination is typical of the modern era, but in mathematics, ideas have been travelling for a very long time.

Not forgetting people

In this chapter I have described how some of the mathematics of the past has survived, in however fragmented a form, over long periods of time and sometimes over long distances too. I have tried to be careful, however, with language. A common word for the passing on of mathematical ideas is 'transmission', but I dislike it: apart from the connotation of radio masts, it suggests that the originators were deliberately aiming their ideas and discoveries at future generations. This has rarely been the case. For the most part, mathematics is written down for one's own use or for one's immediate contemporaries, and its survival much beyond that is largely a matter of circumstance. I have also tried to avoid speaking of ideas simply spreading, as though they were garden weeds with a power of their own.

On the contrary, every mathematical exchange, large or small, is brought about by human agency. Behind the long-running stories outlined above lie innumerable tiny interactions and transactions. We have already glimpsed some of them: Indian envoys presenting themselves to the caliph in Baghdad; a Byzantine scribe copying a manuscript he may barely have understood;

Florentine traders haggling in the markets of Alexandria; a library-keeper in that same city of Alexandria a millennium earlier carefully listing the scrolls in his charge, and perhaps, like John Leland later, distraught at the thought of their destruction; Fermat sending his letters in false hope to Wallis in Oxford; Wiles delivering first news of his proof in a lecture, and news of its eventual correction by email. Mathematical ideas move around only because people think about them, discuss them with others, write them down, and preserve relevant documents. Without people, there is no dissemination of mathematical ideas at all.

Chapter 4
Learning mathematics

A fact easy to overlook is that the largest group of people in modern society doing mathematics is composed not of adults but of schoolchildren. A young person anywhere in the world who is fortunate enough to have access to education is likely to spend a significant amount of time learning mathematics; in the developed countries, this is likely to amount to two or three hours of each school week for ten years or more.

In the light of that, it is a little surprising to remember that the inclusion of mathematics in the school curriculum is a modern phenomenon. Around 1630, for example, John Wallis, later Savilian Professor of Geometry at Oxford, learned arithmetic neither at school nor at Cambridge but from his younger brother studying to go into trade; 30 years later, the highly intelligent and literate Samuel Pepys, also educated at Cambridge, and a member of the Navy Board, struggled to learn his multiplication tables. Nevertheless, passing on mathematical knowledge to at least a few of the next generation has been regarded as an important task in most civilized societies.

A study of what has been taught, and how, tells us a good deal about what aspects of mathematics have been regarded as relevant, and for what purposes. In this chapter we will examine two case studies for which we have relatively good documentation:

a schoolroom at Nippur in southern Iraq some time before 1740 BC, and another at Greenrow Academy in Cumbria in the north of England shortly after AD 1800.

A Babylonian schoolroom

The ancient city of Nippur, situated in the marshlands of the Euphrates about halfway between the modern cities of Baghdad and Basra, was an important religious centre, built around a temple complex dedicated to the god Enlil. Like the abbeys and monasteries of medieval Europe later, Babylonian temples received substantial offerings, and controlled land and labour, and so needed trained scribes who could handle written accounts and calculations. Children destined for the profession, which usually ran in families, probably began their training early.

A small mud-brick house in Nippur, now known as House F, appears to have been one of perhaps several scribal schools in the city. Close to a temple dedicated to the goddess Inana, House F was first built some time after 1900 BC and was used as a school shortly before 1740 BC. Like all mud-brick structures, it required regular upkeep and after it ceased to be used as a school it was rebuilt for the fourth or fifth time. At this point, the builders made good use of hundreds of discarded school tablets which they incorporated into the floor, walls, and furniture of the new house. Other partially destroyed tablets have been found mixed with large quantities of unused clay in recycling bins.

When used as a school, House F was divided into three or four inner rooms and two courtyards, the latter containing benches and the recycling bins. Unfortunately, we do not know the names or ages of the students, of whom there may have been no more than one or two at a time, nor do we know how often or for how long they occupied the courtyard benches. Remarkably, however, their method of using their tablets has enabled cuneiformists to reconstruct their curriculum.

Many of the tablets from House F are flat on one side (the obverse) and slightly rounded on the other (the reverse). The obverse contains on the left a model text written by a teacher, copied on the right by the student. The rounded reverse of the tablet, however, contains longer passages of material learned earlier, rewritten for further practice or perhaps as a test of memory. From some 1,500 Nippur tablets of this kind, each containing 'earlier' and 'later' material, Niek Veldhuis in the 1990s was able to discern a consistent order in the elementary curriculum, starting with basic writing techniques and ending with the beginnings of literary Sumerian. Applying the same methodology to about 250 similar tablets from House F, Eleanor Robson was able to do the same for the House F curriculum, and so to discover the place of mathematics within it.

The student's first steps were to learn the correct technique for writing cuneiform signs and to combine them to form personal names. Then they acquired written vocabulary through lists of words, beginning with trees and wooden objects, then reeds, vessels, leather and metal objects; animals and meats; stones, plants, fish, birds, and garments; and so on. Some mathematical vocabulary was already introduced here, with measures of capacity for boats; of weight for trees and stones; and of lengths for reed measuring-rods. Further metrological units also appeared in dedicated lists of weights and measures later.

Next, the student was expected to learn by heart lists of inverses (pairs of numbers that multiply to 60) and more than 20 standard multiplication tables. A list of inverses, for example, might begin

2	30
3	20
4	15
5	12
6	10
8	7 30
9	6 40
10	6
12	5

(In sexagesimal arithmetic, which we still use for hours, minutes, and seconds, 7 30 is equivalent to $7\frac{1}{2}$, and 6 40 to $6\frac{2}{3}$.)
Multiplication tables required considerable feats of memory. The multiplication table for 16 40, for example, began

1	16 40
2	33 20
3	50
4	1 06 40
5	1 23 20

It has been estimated that it could take up to a year to learn the full set of tables alongside other school exercises. At this stage, students also started to write complete sentences in Sumerian, some of them containing the metrological units learned earlier.

Only after all this, as they also learned more advanced Sumerian, did students begin to carry out their own calculations of reciprocals or inverses not in the standard tables. One of the few 'advanced' tablets from House F contains some of the calculations used to find the inverse of 17 46 40 (answer: 3 22 30). These are written on the same tablet as an extract from a literary work known as 'The supervisor's advice to a younger scribe', which includes some moralistic lines based on the supervisor's memory of himself as a young student:

Like a springing reed, I leapt up and put myself to work.
I did not depart from my teacher's instructions;
I did not start doing things on my own initiative.
My mentor was delighted with my work on the assignment.

Most of the advanced texts from House F are not mathematical
but, like 'The supervisor's advice', literary compositions. Many,
however, contain references to the uses of both literacy and
numeracy in the just administration of society. Lines from a hymn
to Nisaba, the patron goddess of scribes, praise her for bestowing
her gifts upon the king:

A 1-rod reed and a measuring rope of lapis lazuli,
A yardstick, and a writing board which gives wisdom.

A Cumbrian schoolroom

Greenrow Academy was founded in 1780 by John Drape, at Silloth
on the north-west coast of England, just a few miles south of the
Scottish border. Like the school in House F at Nippur, Greenrow
Academy was something of a family enterprise. Drape's father,
known as John Draper, had previously run a school in
Whitehaven, 30 miles south on the same coast. The Whitehaven
school had emphasized subjects relevant to 'trade and seamanship'
and Draper had published two textbooks for the use of his pupils:
*The Young Student's Pocket Companion, or Arithmetic, Geometry,
Trigonometry, and Mensuration, Calculated for the Improvement
of Youth at School* (1772) and *The Navigator's veni-mecum: or a
Complete System of the Art of Navigation* (1773). When Draper
died in 1776, his son John inherited his books, his mathematical
instruments, and some property, enabling him to establish
Greenrow Academy a few years later. After Drape himself died in
1795, the school passed into the care of another family member,
Joseph Saul, a relative of Drape's wife, who remained in charge of
it for nearly 50 years. The curriculum was broadened to include

Greek, Spanish, and scriptural studies, but Greenrow Academy, like its parent school in Whitehaven, continued to place a strong emphasis on mathematical studies.

The school attracted boys not only from the local area but from elsewhere in England and even from overseas. Nine-year-olds, and at one time even a six-year-old, could be registered, but young men in their early twenties were also sometimes educated there. Most of the pupils, however, were aged about 14 or 15. Records for 1809 show that one of the youngest pupils, Rowland Cowper (aged 11), and one of the oldest, James Irving (aged 23), followed the same basic curriculum of English, writing, and arithmetic. Most of the other boys also studied drawing and either French or Latin together with a wide range of mathematical topics. The curriculum followed by John Coleman (aged 15) was typical: English, French, writing, drawing, arithmetic, geometry, trigonometry, mensuration, surveying, book-keeping, spherics, astronomy, mechanics, algebra, and Euclid. Further mathematical subjects on offer were dialling (the construction of sun-dials), gauging, and fortification, while George Peat (aged 16), who seems to have been exceptionally able, also took lessons on conic sections and fluxions (the Newtonian version of the calculus).

We are fortunate, however, to have more from Greenrow than mere lists of topics. Before his death in 2005, the mathematics educator John Hersee collected over 200 mathematical copy books written by pupils in schools throughout England and Wales between 1704 and 1907. These were not exercise books in the modern sense. Pupils did not waste precious paper practising similar questions over and over again; instead they inscribed carefully copied examples of standard problems, thus creating for themselves a collection of worked examples which they could carry with them into later life. Many of the examples were taken from popular textbooks of the time, in particular from the long-running *The Tutor's Assistant* of Francis Walkingame (first published in

1751), but many others must have been invented for their pupils by teachers themselves.

The Hersee collection includes five mathematical workbooks by Robert Smith from the years 1832 and 1833 (see Figure 4). Over these two years Robert filled almost 1,700 pages with mathematical examples, so that we have a very detailed picture of

4. The front page of Robert Smith's arithmetic book, Greenrow Academy, 1832

what he studied. These books were not the first that Robert wrote because he had already advanced beyond the elementary operations of addition, subtraction, multiplication, and division. The earliest surviving book, for 1832, begins with the Rule of Three. This was the rule that enabled countless generations of students to answer questions like: *A* men dig a ditch in *B* days, how long would it take *C* men to do the same job? The rule is so named because there are three known quantities (A, B, C) from which a fourth (the answer) must be found. The rule originated in India and probably travelled westwards with the Indian numerals: it was ubiquitous in Islamic and European arithmetic texts for centuries.

The Rule of Three was taught by rote: like his Babylonian predecessors, a 19th-century English schoolboy was not expected to 'start doing things on my own initiative'. In the example above, he would be taught that he must multiply *B* by *A* and divide by *C* to get the correct answer. But of course, there were always variations to catch out the unwary student: Robert Smith had to learn the Rule of Three Direct, the Rule of Three Inverse, and the Double Rule of Three. These topics were followed by, among others, Barter, Interest, the Rule of Fellowship (the sharing of profits), Vulgar Fractions, Decimal Fractions, and Arithmetic and Geometric Progressions. A second book, apparently written in the same year, works through a similar list of topics, again beginning with the Rule of Three and ending with Progressions and Duodecimals. The books seem to have been written consecutively, since Robert himself labelled them Vol. I and Vol. II, and it is not clear why he worked through similar material twice.

Many of his examples are taken from Walkingame; here, for instance, is one of just two examples on Permutations (the other is on the number of changes that can be rung on 12 bells):

A young person, coming to town for the convenience of a good library, made a bargain with the person with whom he lodged, to give him £40 for his board and lodging, during so long a time as he could place the family (consisting of 6 persons besides himself) in different positions, every day at dinner. How long might he stay for his £40?

Robert wrote the correct solution (1 × 2 × 3 × 4 × 5 × 6 × 7 = 5040 days) immediately after the question, but then, following Walkingame quite closely at this point, moved on immediately to vulgar fractions.

Robert's two arithmetic books for 1832 between them contain almost 900 pages. In addition, he filled almost 500 pages more in a third book headed 'Geometry Trigonometry Mensuration and Surveying', which contains some of the beautifully drawn and painted sketches that seem to have been encouraged at Greenrow (see Figure 5).

The next book, with the title page 'Arithmetic by Robert Smith Green-Row 1833', is on 'Practical questions upon common rules'. The questions known as Bills of Parcels are of particular interest because pupils often substituted their own names and dates for those given by Walkingame. Thus Robert's first Bill begins:

Green Row July 13th 1832
Mr Thos Nash
Bought of Robert S Smith

| 8 Pair of worsted stockings | at 4s 6d per pair | £1 16s 0d |
| 5 Pair of thread ditto | at 3s 2d | 15s 10d |

Further dates on further bills follow through July 1832 and into August, suggesting that Robert might have been writing this book in 1832, and did not begin it but only finished it in 1833, the date on the title page. The name of Thomas Nash is elsewhere inscribed at the end of Robert's first book, along with that of one

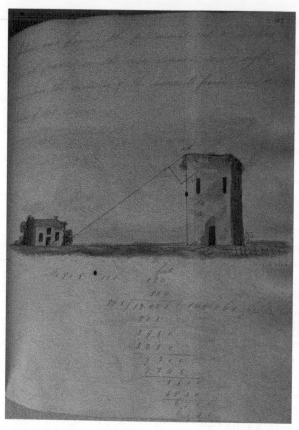

5. A problem in trigonometry, illustrated and answered by Robert Smith, Greenrow Academy, 1832

Robert Reid, suggesting that they might have been his teachers; Robert Reid makes another appearance here, charged with

| 18 Yards of fine lace | at £0 12s 3d per yard | £11 0s 6d |
| 5 Pair of fine kid gloves | at 2s 3d per pair | 11s 3d |

and so on.

The second book completed in 1833 was on 'Mensuration of Solids'. It includes sophisticated calculations of the volumes and surface areas of the five regular solids (tetrahedron, cube, octahedron, dodecahedron, icosahedron) but also calculations typical of those used by bricklayers, masons, carpenters, slaters, painters, glaziers, plumbers, and others, with the appropriate units used by each. Robert learned that painters, for example, estimate the areas of 'wainscottings, doors, window shutters' by the square yard but that 'deductions must be always made, for fireplaces and other openings'.

Unfortunately, we do not know how old Robert was when he did all this, but we can see that his years at Greenrow gave him a mathematical education that was both theoretical and thoroughly practical.

Girls

I have hesitated to include a section that treats a group of people who constitute half of humanity as though they were a minority, but there is no getting away from the fact that for most of history in most societies it has not been thought necessary, or indeed appropriate, to educate girls, and certainly not in subjects like mathematics or science. It is not surprising, therefore, that there have been very few women mathematicians of note, just as until recent centuries there have been few women writers, lawyers, or doctors. This state of affairs must have left countless thousands of intelligent women somewhat frustrated. Nevertheless, there have from time to time been some who were given, or who created for themselves, opportunities for mathematical education.

One group of such women were those with enough wealth or leisure to indulge in whatever studies they chose. An early example was the Chinese Empress Dèng 鄧, who at the end of the 1st century AD took lessons in *suàn shù*. Unusually for this period, her teacher was also a woman, by the name of Bān Zhāo 班昭. Much later, in the 1640s, Princess Elisabeth of Bohemia and Queen Christina of Sweden both took lessons from Descartes, though they were perhaps more interested in his philosophy than in his mathematics. A century later, Europe's most prolific mathematician, Leonhard Euler, wrote over 200 letters on mathematical and scientific subjects to the Princess of Anhalt-Dessau, niece of Frederick the Great of Prussia. The letters, published in French, Russian, and German, and later in English as *Letters to a German Princess*, remain in print to this day.

A more common route into mathematics for ordinary women, however, was to be taught by a father, husband, or brother. In the 19th century BC, for example, there were two women scribes in the Babylonian town of Sippur, the sisters Inana-amaǧa and Niǧ-Nanna. It seems more than likely that they learned the profession from their father, Abba-ṭābum, also a scribe. Two millennia later, the Empress Dèng and her brothers received their earliest education from their father, even though their mother, it seems, thought this a waste of time for a girl. The Empress Dèng's later teacher, Bān Zhāo, was the sister of the scholar Bān Gù 班固, whose work she understood sufficiently well to complete after his death, including a treatise on astrology. Perhaps the most famous father–daughter pair in mathematics is that of Theon and Hypatia in late 4th-century Alexandria; from Hypatia herself, however, we have no writings, only secondary accounts of her life and death around which much legend has accumulated.

Teaching of girls within the family continued into the early modern period. John Aubrey, writing in the 1670s about his former friend Edward Davenant, vicar of Gillingham in Dorset, noted his love of mathematics, though 'being a Divine he was unwilling to print, because the world should not know how he spent a greater part of his time'. Davenant had taught algebra not only to Aubrey himself, but also to his own daughters:

He was very ready to teach and instruct. He did me the favour to informe me first in Algebra. His daughters were Algebrists.

As it happens, we know what Edward Davenant taught his eldest daughter, Anne, in the way of algebra because in 1659 Aubrey, avid recorder of all human affairs, copied out the contents of her notebook. Anne was born before 1632 (the date of birth of her younger sister Katherine) and married Anthony Ettrick in 1650, so was probably being trained as an 'Algebrist' some time in the early 1640s. Aubrey's copy of her work is headed

This algebra I transcribed from ye MS of M^{ris} Anne Ettrick the eldest daughter of Dr Davenant who is a very good Logist.

The early problems in Anne's notebook, and the Latin in which she wrote them, are typically those of a young beginner. In one of them, for example, some girls are taking a walk when a young man comes along. 'Greetings', he says (in Latin), 'you twelve maidens'. To which one of the girls replies immediately (and also in Latin): 'If we were multiplied by five, we would be as many more than twelve as we are now fewer than twelve.' Question to the reader: how many girls were there? Several pages later, we find Anne working an example whose form and solution had been proposed by al-Khwārizmī in Baghdad eight centuries earlier: what number multiplied by 6 and added to 16 makes its own square (in modern notation $6x + 16 = x^2$). Finally, towards the end of the notebook, both the Latin and the mathematics become more mature. The penultimate problem comes straight from the *Arithmetica* of Diophantus: to divide 370 into two cubes, whose roots are whole numbers that add to 10. As Anne was able to show, the answer is 7^3 plus 3^3. The numbers in this question have been carefully chosen to give an easy solution, but it becomes impossible if 370 is replaced by a perfect cube, as Fermat, at much the same time but far away in Toulouse, was just discovering.

Well into the 18th century, girls were likely to be taught mathematics only if, like the Empress Dèng or Anne Davenant, they had the advantages of social status or indulgent parents. Sophie Germain, one of the key people to make progress on Fermat's Last Theorem, benefited from both. Born into a wealthy and educated family in Paris in 1776, she was just 13 when the French Revolution broke out. Confined to her home, she amused herself in her father's library and discovered mathematics, a subject her parents did not at first think suitable for her but they later relented in the face of her determination. When she was 18, she managed to obtain lecture notes from the newly opened École Polytechnique and, although she was not allowed to attend lectures herself, she submitted work under the pseudonym Monsieur le Blanc to one of the greatest of the École's teachers, Joseph-Louis Lagrange. Four years later, she corresponded with the equally great German mathematician Carl Friedrich Gauss, again under the pseudonym le Blanc. To the credit of both Lagrange and Gauss, they continued to admire her mathematics and her courage even after they discovered her true identity. Sophie struggled against the odds for most of her life: she had never had the kind of education that a boy of equal talent might have had, and her work was often marred by errors and incompleteness. She never held any official post. Nevertheless, after her death in 1831 Gauss remarked that she would have been worthy of an honorary degree at Göttingen, by then one of the key mathematical centres of Europe.

Articles or posters about 'Women in mathematics' invariably feature Hypatia and Sophie Germain, unfortunately not because they were typical of their time and place but because they were not. Unsung women like Bān Zhāo or Anne Davenant are on the whole much more representative of women's experience of mathematics and mathematical education.

By the 19th century, the situation for girls in western Europe slowly improved as they started to benefit in larger numbers from elementary school education. Copy books by girls are in the minority in the Hersee collection, but those that we have give us some valuable insights into the kind of mathematics taught to girls in a variety of schools in England and Wales.

In 1831, the year before Robert Smith at Greenrow began creating the books discussed above, Eleanor Alexander, at Fairwater School in a valley north of Newport in South Wales, worked on Reduction ('Bring £30 1s $1\frac{1}{4}$ into farthings') and the Rule of Three ('If 7 yards of cloth cost £3 10s what will 65 yards cost?'). Her entire copy book of 127 pages consists of just these two types of question. Three years later, beginning in October 1834, Ann Weetman at Appleton-le-Moors, near York, began to work her way through Walkingame's *The Tutor's Assistant* (see Figure 6). Her early entries are dated, so we know that she spent about ten days on Simple Addition but a whole month on Multiplication. After Christmas she worked on Compound Addition (of money, cloth measure, land measure, beer and ale measure, and so on) and by the end of March she reached Bills of Parcels. In her case, the 8 pairs of worsted stockings went to Mrs Wm G. Atkinson (her teacher?) while Mr Henry Weetman (her father? her brother?) bought 15 yards of satin. By April 1837 she had reached Practice, a method that relied on knowing fractions of standard weights and measures (see Figure 7). Her book ends after 250 pages on 10 May 1837, by which time she had worked her way through Walkingame as far as Compound Interest, considerably less in three years than Robert Smith had completed in a few months, but still a respectable amount of mathematics.

Twenty years later, Elisabeth Attersall in Stainfield in Lincolnshire also worked through Walkingame, from Compound Addition to the Rule of Three ('Gave £1 1s 8d for 3lb of coffee, what what [sic] must be given for 29lb 4oz.'). In her case, the 8 pairs of worsted stockings went to Mrs Chappell on 22 October 1850. Miss I. Norman at Mr Ingleson's Dorset Street Academy in Hulme, Manchester, in 1861, however, was served a slightly

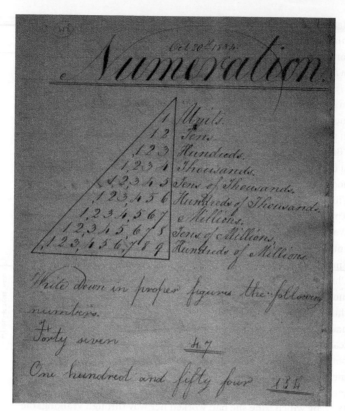

Oct 20th 1834

Numeration.

1	Units.
1 2	Tens.
1 2 3	Hundreds.
1,2 3 4	Thousands.
1 2,3 4 5	Tens of Thousands.
1 2 3,4 5 6	Hundreds of Thousands.
1,2 3 4,5 6 7	Millions.
1 2,3 4 5 6,7 8	Tens of Millions.
1 2 3,4 5 6,7 8 9	Hundreds of Millions.

Write down in proper figures the following numbers.

Forty seven   47

One hundred and fifty four   154

6. The first page of Ann Weetman's exercise book, dated 'Oct 20th 1834'

different diet. Her copy book was specially printed with the name of the school, on pale blue paper with margins ruled in double red lines. On the first page, Miss Norman wrote 'Progressive arithmetic' but sadly she did not progress very far: every one of the book's 60 pages is filled with multiplication or division of pounds, shillings, and pence ('What must I pay for 4767 yards of cloth at $7\frac{8}{9}$d per yard?'). Elizabeth Dawson at Carshield School in Northumberland a year later did a great deal of work on the Rule of Three and a great deal more on Practice; to find 'the value of

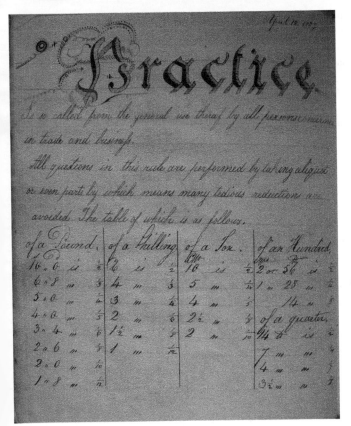

7. **One of the final pages of Ann Weetman's exercise book, dated 'April 12 1837'**

7234 yards at 6s 8d', for example, she used a fact known to every British schoolchild before 1971, that 6s 8d was £$\frac{1}{3}$. It was rather less easy, though, for her to find the cost of 65 feet 3$\frac{1}{4}$ inches at 3s 7$\frac{3}{4}$d per foot.

From April 1866, Isabella Lund, a pupil at the Grammar School in Bolton-le-Sands in Lancashire (founded originally only for boys),

spent just over a year progressing from Simple Addition to the
Rule of Three. A year later, Miss G. Jones at Ribston Hall
Grammar School for Girls in Gloucester gradually filled out 20
pages of invoices; her 8 pairs of worsted stockings went to a Miss
Jenkins in July 1868.

The girls' copy books selected above are some for which we
happen to know the name of the writer and the school and
the date; without further research we cannot assume they
are representative. They do suggest, however, that
mathematical education for girls had a strong practical emphasis
(no Euclid here); further, by modern standards, the pace was
sometimes agonizingly slow and repetitive. Nevertheless, the girls
who wrote these books were both literate and numerate,
particularly compared with their counterparts in previous
generations.

To go beyond elementary mathematics to university education,
however, still required a particular strength of character. We will
end this section with a comparison of two women who did manage
to reach the higher reaches of their country's education systems,
Flora Philip in Scotland and Florentia Fountoukli in Greece.

Flora Philip was one of the first women graduates of Edinburgh
University, in 1893, but she had already joined the Edinburgh
Mathematical Society seven years earlier. Most of her education in
higher mathematics was gained in fact not from the university but
through the Edinburgh Association for the University Education
of Women. The Association had been established in 1867 to
provide education beyond school level for women, parallel to that
provided by the university for men. Mathematics was included in
the Association's lecture courses early on, though not without
some opposition from those who considered it 'altogether outside
the domain of a lady's thought'. The aim was to teach the same
mathematics as was taught at the university, but since many of the
women were ill-prepared by their elementary schooling, the level

reached was never as high as in the university courses. Nevertheless, they were taught Euclid, algebra, trigonometry, and conic sections. The number of women who took the courses was often very low but, as one of the lecturers reported, 'The zeal and industry of the Class doubly compensates for the smallness of its numbers'. Later, a more advanced senior course was introduced, from which Flora Philip graduated successfully in 1886, the year she joined the Edinburgh Mathematical Society. By the time she was awarded her degree from the university in 1893, she had already been teaching for some time at St George's School for Girls, a school founded by the Association. In that same year, however, she married, and thereafter withdrew both from academic life and from the Edinburgh Mathematical Society.

The career of Florentia Fountoukli, born in 1869 in Athens, ran parallel in several ways to Flora's. While Flora was studying mathematics with the Edinburgh Association in the 1880s, Florentia Fountoukli was earning her schoolteacher's diploma from the Arsakeion Normal School for girls in Athens. Afterwards the Board of the Arsakeion Schools allowed Florentia funds to study pedagogy in Berlin for a year. She then requested an extension in order to take a degree in Zürich, in mathematics, but the Board refused. (Her brother Michael, on the other hand, did become a mathematician and later worked in Hamburg.) Florentia returned to teach in the Arsakeion School in Corfu, during the same years that Flora was teaching at St George's. In 1892, just as Flora matriculated into the University of Edinburgh, Florentia matriculated into the mathematics department of the University of Athens, the first woman to do so. Unlike Flora, however, she does not appear to have graduated. Instead she continued to teach, in a school for girls in Athens, which she and her friend Irene Prinari founded and ran themselves. By 1899, she was signing herself Fountoukli-Spinelli which suggests that she may have married Loudovikos Spinelli, another teacher, but the facts are not clear. Unfortunately, by the late 1890s, before she was even 30, her health began to fail and she went to live in Italy, where she died in 1915.

Both Flora and Florentia had to struggle to get the kind of education they wanted. Nevertheless, the universities of Edinburgh and Athens were ahead of some others. Cambridge University did not admit women as full members until 1947.

Autodidacts

Until two centuries ago, only a small number of girls anywhere in the world received any kind of mathematical education. But even for boys, compulsory education in mathematics has been a relatively recent phenomenon. In 17th-century England, as we saw with Wallis and Pepys, it was possible to be educated all the way through school and university without learning much at all in the way of mathematics. Those with a particular aptitude or liking for the subject were therefore often largely self-taught. This was the case for Fermat, who learned some of the most advanced mathematics of his day from the books owned by the father of his friend Etienne d'Espagnet in Bordeaux. It was also the case for Isaac Newton, one of the greatest mathematicians of the 17th century. Newton may have learned some elementary mathematics at his grammar school in Grantham in Lincolnshire, but he learned very much more through his own reading as a student at Cambridge in the 1660s. Many years later, he described to a friend how he had read Descartes' *Géométrie*, republished in Latin just a few years earlier. Many people will recognize the difficulties of reading a strange new mathematical text; perhaps rather fewer will have emulated Newton's stubborn and self-motivated persistence.

> He bought Descartes's Geometry & read it by himself. When he was got over 2 or 3 pages he could understand no farther, then he began again & got 3 or 4 pages farther till he came to another difficult place, then he began again & advanced farther & continued so doing till he made himself Master of the whole.

We know from Newton's surviving manuscripts that he proceeded in a similar way with other contemporary texts, and that he worked on the material he found in them to create mathematics that went far beyond that of any of his predecessors.

In the 17th century, and some way into the 18th, a person sufficiently motivated could still read and learn from almost the entire range of available mathematical literature. Even at the beginning of the 19th century, Sophie Germain managed to teach herself some of the most advanced mathematics of her day, but she belonged to almost the last generation for whom this was feasible. By the 20th century, it was barely possible except, perhaps, for prodigies with quite exceptional mathematical instincts, like the self-taught Indian mathematician Ramanujan. Andrew Wiles was certainly not self-taught; he went through the many years of formal education that are now needed to initiate even the most mathematically gifted into some of the problems, techniques, and conventions of the discipline. The days of 'amateur' mathematics when almost anyone could have a go at a cutting-edge problem like Fermat's Last Theorem are long past.

Nevertheless, from time to time writers continue to create fictional accounts of individuals who manage to learn mathematics from the writings of another person, well enough to understand and extend their work. One such story is *Mrs Einstein* by Anna McGrail, another is the more recent play *Proof* by David Auburn. In both, the heroines are the daughters of mathematicians (a trope we have already seen in real life) who manage to educate themselves to very high levels in their fathers' work. Unfortunately, the reality of modern mathematics is that such feats are now extremely implausible.

Why learn mathematics at all?

Given the enormous amount of human energy that over the centuries has gone into teaching or learning mathematics, it seems a little perverse to ask 'Why?' The answers to the question, however, have varied considerably over time. Sumerian texts of the 2nd millennium BC made it clear that literacy and numeracy were essential to the just administration of society, though that may have seemed a somewhat distant ideal to the boys on the cramped courtyard benches of House F.

Two thousand years later, boys of a similar age educated in the abacus schools of 13th-century Italy learned, like their earlier Babylonian counterparts, to handle numbers, weights, and measures, but for different reasons: not for the good of society as a whole but so that they as individuals would be better fitted for the commercial ventures they were expected to engage in. The value of mathematical skills to the individual is seen again in Robert Recorde's preface to *The Pathway to Knowledg* with its long list of specific crafts and occupations that require a knowledge of geometry.

In Recorde's writings, however, we also get a glimpse of yet another reason for studying mathematics: to whet the wits, that is, to sharpen the mind. Recorde was not the first to suggest this. Some mathematical puzzle questions ascribed to the 8th-century teacher Alcuin were entitled 'Propositions of Alcuin for sharpening the young' (*'Propositiones Alcuini ad acuendos juvenes'*). The idea that mathematics should be learned, like Latin or Greek, to improve one's mind has continued ever since. After all, the mathematics required for ordinary daily life, basically time-keeping and accounting, has probably been acquired by most people by the end of childhood. Few adults ever need to use Pythagoras' Theorem, or solve quadratic equations, or bisect an angle, but almost all were once taught to do so. It can be argued (and I would argue) that learning a foreign language or studying history does just as much to encourage the development of memory, reasoning, and analysis, but such subjects have never acquired the prestige of mathematics, and at present are optional rather than compulsory subjects in the British school curriculum.

Perhaps it is the sheer longevity of mathematics that has made it such an integral part of every modern child's education. It is also the case that those who are to progress to the frontiers of the subject need, like young musicians, to start young and practise regularly.

Chapter 5
Mathematical livelihoods

Any mathematician who wants to break new ground needs time to think and scribble, and some kind of financial support. Let us return for a moment to those we met in the first chapter. We have no idea how Diophantus earned his living; perhaps, like many with mathematical talent, he taught. Many of the best-known mathematicians of the century before Fermat also taught mathematics, but often only as a secondary occupation: Girolamo Cardano and Robert Recorde were physicians, though Recorde also worked for much of his life in mints and mines; Rafael Bombelli and Simon Stevin were both employed on practical construction projects; François Viète, like Fermat, was a lawyer and Counselor. Fermat has often been described as an 'amateur' mathematician, but he lived at a time when there were so few professionals that the concept of amateur was meaningless. Wiles, on the other hand, cannot be described as anything but a professional, fully accredited, and paid to work full time in the research and teaching of mathematics.

Over the centuries, there have been significant changes in the ways that mathematicians have been employed. A modern mathematician is very likely to work in education, finance, or industry, all of which are institutionally organized. Some individuals may also be prepared to pay for mathematical services, for tuition, perhaps, or accounting skills, but they keep no more

than a small number of people in work. In the 1st millennium AD
the picture was very different. Economic and political power
throughout most of Europe and Asia was concentrated in the
hands of kings, bishops, caliphs, and warlords. Those who wanted
to live by their intellectual skills, including mathematics, were
wise to place themselves under a patron powerful enough to pay
and protect them. Such patronage could take many different
forms. In this chapter we will see it at work first in the lives of
three scholars of the 10th and 11th centuries in the lands then
dominated by Islam.

Patterns of patronage

Thābit ibn Qurra was born in AD 826 in the town of Ḥarrān, very
close to the modern Turkish–Syrian border, and spent his early life
there as a money changer. He was not a Muslim but belonged to a
local sect, the Sabians. Only a few years previously, the library
known as the Bayt al-Ḥikma (House of Wisdom) had been
established in Baghdad by the Abbasid caliph al-Ma'mūn for the
purpose of preserving, and translating into Arabic, texts in Greek,
Sanskrit, or Persian. Ibn Qurra's knowledge of Greek and Arabic
besides his native language of Syriac brought him to the attention
of the Baghdad mathematician Muḥammad ibn Mūsā when the
latter passed through Ḥarrān on his return from Byzantium.
Unfortunately, we do not know the date of this meeting, but we
may suppose that ibn Qurra was still relatively young because at
ibn Mūsā's invitation he moved to Baghdad, where he was
educated by ibn Mūsā and his two brothers (collectively known as
the Banū Mūsā) in mathematics and astronomy.

In the years that followed, ibn Qurra became one of the most
respected scholars in Baghdad. He wrote on medicine, philosophy,
and religion, but is best remembered now for his work in
mathematics and astronomy. He translated several treatises of
Archimedes into Arabic, and also wrote extensively on topics that
had interested Archimedes: mechanics, and problems to do with

areas, surfaces, or volumes of curved shapes. He commented on Ptolemy's *Almagest* and wrote on spherical geometry and astronomy, especially on the motion and apparent height of the Sun, and the motion of the Moon and the five then known planets. He also studied Euclid's *Elements* intensively; his attempted proof of one of Euclid's postulates, on parallel lines, was taken up again in Oxford in the 17th century. Ibn Qurra also produced his own proofs of Pythagoras' Theorem, one of which is shown in Figure 8.

Ibn Qurra remained in Baghdad until his death in AD 901. He retained his association with the Banū Mūsā for many years and taught ibn Mūsā's sons. During the final ten years of his life, he became a regular attendant at the court of the caliph al-Muʻtaḍid, so intimate with the caliph that, according to a biographical sketch by the 12th-century writer, al-Qiftī, he was allowed to 'sit down in his presence at any time he wished'. Later his son, Sinān, and two grandsons became well known scholars in their own right. In what is known of ibn Qurra's life, we may thus discern two crucial features. One is a network of teaching and learning established between friends and families, in this case linking members of ibn

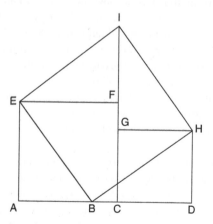

8. Thābit ibn Qurra's proof of Pythagoras' Theorem: a simple cut-and-paste argument shows that IHBE = EFCA + GHDC

Mūsā's family to ibn Qurra's. Such close personal relationships have now been observed several times in the course of this book. The second feature, more particular to ibn Qurra's time and place, is the protection and patronage offered first by the Banū Mūsā, later by the caliph himself.

Another scholar, Abū Rayḥān al-Bīrūnī, generally known as al-Bīrūnī, was born 70 years after ibn Qurra died, at the further end of the Islamic domains, and into a less stable world. The town of his birth on the Āmū Daryā (or River Oxus) lies just inside modern Uzbekistan and is now called Biruni. He was taught by the mathematician and astronomer Abū Naṣr Manṣūr, with whom he continued to work later in his life. As a young man he was already using solar observations to calculate latitudes of local towns, but this activity was disrupted when civil war broke out in 995 and he was forced to flee. We know something of his wide-ranging movements over the next 30 years from his precise observations of eclipses. At times, he worked in the region south of the Caspian Sea, close to modern Tehran, where he is known to have dedicated a text on chronology to the Ziyārid ruler of the region, Qābūs. At other times he lived in his home area, at first under the patronage of the Sāmānid ruler Manṣūr II, later, for 14 years, under that of Abu'l-Abbās Ma'mūn.

This last relatively stable period came to an end in 1017 when the region was overrun by the Ghaznavid dynasty based in what is now eastern Afghanistan. Al-Bīrūnī appears to have been taken prisoner, and subsequently lived for many years near Kabul or at Ghazna itself, about 100km to the south. His relationship to the sultan, Maḥmūd, is unclear: he complained of ill treatment but was later supported in some of his studies. He was also able to travel to northern India, which had also fallen under Ghaznavid rule, and wrote extensively about the region and its religion, customs, and geography. After Maḥmūd's death in 1030, al-Bīrūnī came under a second Ghaznavid patron, Maḥmūd's son Ma'sūd;

and under a third, Ma'sūd's son Mawdūd, after Ma'sūd was murdered in 1040. Al-Bīrūnī himself died in Ghazna in 1050.

Throughout a life beset by dynastic changes, al-Bīrūnī was a committed scholar and a prolific writer. Around half of his works were on astronomy and astrology, with other texts on mathematics, geography, medicine, history, and literature. Unfortunately, only a fraction of what he wrote has survived.

The third mathematical scholar we will look at is 'Umar ibn Ibrāhīm al-Nīsābūrī al-Khayyāmī, better known in the west as Omar Khayyam. He was born shortly before al-Bīrūnī died, in Nishapur in north-east Iran. His name suggests that he came from a family of tentmakers. By this time, the Iranian region had come under the rule of the Seljuqs, a dynasty of Turkic origin. As a young man al-Khayyāmī travelled east to Samarkand where he wrote an important treatise on equations, dedicated to the chief justice, Abū Ṭāhir. He later spent many years in Isfahan where, under the patronage of the sultan Malik-shāh and his vizier Niẓām al-Mulk, he supervised the observatory and the compilation of astronomical tables. During this same period, like ibn Qurra before him, he also wrote commentaries on Euclid. Unfortunately, the observatory was closed in 1092 after Niẓām al-Mulk was murdered and Malik-shāh died. Eventually, after further changes of ruler, al-Khayyāmī left Isfahan. After spending some time in Merv, about halfway between Isfahan and Samarkand, he finally returned to Nishapur, where he died in 1131.

I cannot resist including one of his *Rubā'iyyāt* (quatrains), not from the Victorian translation by Edward FitzGerald but from a 1998 translation by Shahriar Shahriari.

> The secrets eternal neither you know nor I
> And answers to the riddle neither you know nor I
> Behind the veil there is much talk about us, why
> When the veil falls, neither you remain nor I.

These three simplified case studies do not begin to tell all that could be said about mathematical practice under the medieval Islamic dynasties, but at least a few general points emerge. One is that, just as a few centuries earlier, Greek mathematical writers were to be found anywhere around the eastern Mediterranean but rarely in Greece itself, so those who wrote mathematics in Arabic were scattered across an even vaster region, from modern Turkey to modern Afghanistan, though not Arabia itself. For this reason, historians prefer to call such writers 'Islamic' rather than 'Arabic' but, as the example of ibn Qurra shows, not all were Muslims, nor were their mathematical writings related to their religious views. Nevertheless, they all lived in societies where the practices and culture of Islam were dominant and so the label is probably better than any other.

A second theme that comes through is the precariousness of scholarship in a world of rapidly changing rulers and dynasties. For a boy or young man of talent to have his mathematical skills recognized and nurtured was already a matter of chance and circumstance, as for both ibn Qurra and al-Bīrūnī. His ability to study or travel thereafter might depend very largely on the favour and financial support of a ruler whose own future might be far from secure. Al-Bīrūnī seems to have been particularly remarkable in enjoying or surviving the attentions of patrons from opposing dynasties. Despite such difficulties, the output of some of these scholars was both prolific and varied. Those who wrote on astronomy and astrology might also write on spherical geometry and trigonometry, or on the *Elements* of Euclid or the works of other Greek writers, or on arithmetic and algebra, or on geography, history, music, philosophy, religion, or literature.

Finally, one might ask what was in such arrangements for the patron? Individual cases differed greatly, indeed there was no single word in Islamic societies for the relationship here described as 'patronage'. As we have already seen in China and Europe, rulers often valued the mathematically adept for their ability to calculate auspicious dates. In some cases, they may also have hoped for long-term, even eternal, benefit from their support for good works. Moreover, owning the services and companionship of the intellectually talented would have been both a source of pleasure and a sign of prestige.

From about the end of the 12th century, scholars were more frequently able to obtain paid positions at endowed teaching institutions, the *madrasas*, and so became less dependent on the whims or preferences of individual rulers. To examine more closely the shift from patronage to professional employment, however, we will now turn to England at a slightly later date.

From patronage to professionalism

In England, the 40 years from 1580 to 1620 were a period of transition, when patronage still existed but where we may also discern the first signs of movement towards publicly accountable paid positions. The careers of Thomas Harriot, William Oughtred, and Henry Briggs illustrate some of the possibilities and opportunities open to the mathematically talented in England at that time.

Thomas Harriot, born in 1560, studied at Oxford in the years 1577 to perhaps 1580. He did not take a degree in mathematics (there was no such thing at the time) but may have learned something of the subject from tutors or his own reading. Better attested is his interest in exploration and navigation, which he seems also to have acquired at Oxford, possibly from the lectures of the adventurer Richard Hakluyt. During the 1580s Harriot came under the patronage of Walter Raleigh, who at the time was greatly interested in the potential colonization of America. In 1585 Harriot sailed to the coast of what is now North Carolina on a voyage financed by Raleigh, a year-long expedition that ended in failure but which enabled Harriot and his friend John White to bring back a great deal of useful information and some beautiful drawings of the people, flora, and fauna of the region. Unfortunately, he also brought back a fondness for tobacco, which was eventually to kill him.

Before the trip, Harriot had been engaged by Raleigh to teach navigation to the sailors, though unfortunately the text he wrote is now lost. On his return, he continued to live under Raleigh's patronage, at first on Raleigh's estates in Ireland (another colonial venture), later at Raleigh's London home, Durham House on the banks of the Thames. It was from the roof of Durham House that Harriot conducted his early experiments on falling bodies, comparing the rates of fall of balls of metal and wax. Harriot continued to be close to Raleigh right up to the day when Raleigh was executed in 1618: notes on Raleigh's final speech from the scaffold survive in Harriot's handwriting amongst his personal and mathematical papers. By the early 1590s, however, Harriot had a second patron in Henry Percy, ninth Earl of Northumberland. Harriot spent the remaining 30 years of his life at Percy's London home, Syon House in Middlesex on the Thames, or at his country home, Petworth House in Sussex. Unfortunately, neither of Harriot's patrons successfully negotiated the political and religious tensions of the day: Percy, like Raleigh, spent many years imprisoned in the Tower of London. Nevertheless, he supplied Harriot with an income and the freedom to pursue whatever studies he chose. Harriot never lost his interest in the problems of navigation at sea; he also later turned to astronomy, and contemporaneously with Galileo used a telescope to observe sunspots and the craters of the Moon. Through one of his Oxford friends, Nathaniel Torporley, he managed to obtain the mathematical works of Viète (which later so profoundly influenced Fermat), so becoming one of the first people anywhere, and certainly the first Englishman, to appreciate and extend some of the exciting new mathematical ideas that were being developed in France.

Harriot published none of his findings. With a secure private income, he had no need either to prove himself or earn a living. Nor did he teach, though he did discuss his ideas within his own circle of friends. In one sense, Harriot's work had little immediate influence; he certainly did not cause the kind of intellectual stir that Galileo later did. On the other hand, his freedom to work as he chose enabled him to explore a wide range of subjects, some of them fairly arcane, and to carry his investigations to some important conclusions. The modern term for this is 'blue skies research'. Harriot's work could easily have been lost, but fortunately his reputation amongst his contemporaries was such that after his death in 1621 his papers were preserved, and some of the ideas in them continued to circulate amongst his successors over many years. In that sense, Harriot can be said to have encouraged, albeit indirectly, both mathematical discussion and the respect for mathematical and scientific studies that characterized the fledgling Royal Society half a century later. Indeed, such was Harriot's reputation that in its first ten years the Royal Society more than once instigated searches for his surviving papers.

Less creative than Harriot but in some ways of equal importance to the later flourishing of mathematics in England was William Oughtred. Born in 1573, Oughtred was only a few years younger than Harriot but outlived him by about 40 years. From 1610 or earlier, Oughtred was a clergyman in Albury in Surrey; he seems never afterwards to have moved away from there apart from occasional visits to London. He became renowned as a teacher of mathematics, to both children and adults. Like Harriot, he also acquired an aristocratic patron, Thomas Howard, Earl of Arundel, whose country seat at West Horsley was only a few miles from Albury. Oughtred taught Howard's son William as he taught the sons of other local gentry. Through Howard, he also met a relative of the family, Sir Charles Cavendish, who proved to be an important figure in English mathematics at this period. Cavendish was not particularly good at mathematics but for some reason was fascinated by it, and avidly collected and tried to understand the latest books and papers. After Harriot's death, for instance, he copied out entire sections from Harriot's manuscripts, though, he admitted, 'I doute I understand not all'. It was Cavendish who brought Oughtred the work of Viète from France just as Torporley had earlier brought it for Harriot.

It was also Cavendish who encouraged Oughtred to write his first textbook, dedicated to his pupil, the 14-year-old William Howard. The book, published early in 1631, became known by its abbreviated title, *Clavis mathematicae* (*The Key to Mathematics*), and it ran and ran, through five Latin editions and two English translations. The content was elementary, an introduction to arithmetic and algebra, but by that time Recorde's earlier textbooks were almost a century old and there was a desperate need for something fresh. When new professors were installed at the University of Oxford after the years of civil war, it so happened that both were pupils or readers of Oughtred and they immediately introduced the *Clavis* to Oxford, making it the first mathematical book to be printed by the university. Almost every 17th-century mathematician of note, and many who were not, took some of their first steps with the *Clavis*, Christopher Wren, Robert Hooke, and Isaac Newton among them. Thus although Oughtred himself never made any great mathematical advances and taught only at a relatively elementary level, he, like Harriot, indirectly encouraged the dissemination and development of mathematical expertise in early modern England.

Neither Harriot nor Oughtred could have done what they did, however, without the support of the three aristocrats who encouraged their work: Henry Percy, Thomas Howard, and Charles Cavendish. A later member of the Cavendish family gave his name to the Cavendish Laboratory in Cambridge, but the Percy and Howard families have not usually been associated with science or mathematics; nevertheless, without the trust and the intellectual and financial support offered by these three men, it might have taken very much longer than it did for a mathematical community of critical size to emerge in England during the first half of the 17th century.

At the same time, and by contrast, we should not overlook certain other contemporary developments. In 1597, a legacy left by the merchant and financier Thomas Gresham paid for seven public lectureships (one for each day of the week) in astronomy, geometry, physic (medicine), law, divinity, rhetoric, and music. Gresham College (which survives to this day and still offers public lectures) played its own role in strengthening the London mathematical community; meetings held after the lectures during the 1650s helped to bring about the establishment of the Royal Society a few years later. Twenty years after the founding of the Gresham lectureships, Henry Savile founded the chairs of geometry and astronomy at Oxford. For many years there was fluid movement between the posts at Gresham and Oxford. In particular, the first Gresham professor of geometry, Henry Briggs, also became the first Savilian professor of geometry at Oxford.

Briggs, from Halifax in Yorkshire, was almost exactly the same age as Harriot and entered St John's College, Cambridge, in 1577, the same year that Harriot was registered at Oxford. Unlike Harriot, however, he followed a university career, lecturing at Cambridge first in medicine, later in mathematics, before he moved to Gresham College in 1597. He was there for over 20 years until he took up the Savilian chair in Oxford, where he remained until his death in 1630.

Briggs and Harriot make a fascinating pair: one of the tantalizing unanswered questions in the history of mathematics of this period is whether they ever met. They ought to have done. During the years before and after 1600, Briggs, like Harriot, was intensely interested in problems of navigation. In 1610, while Harriot was observing sunspots, Briggs was working on eclipses. When John Napier brought out his 'wonderful invention' of logarithms in 1614, both Harriot and Briggs soon became aware of it. Briggs immediately travelled to Scotland to visit Napier, and helped him to develop his work further; Harriot no longer made long journeys and in any case was already becoming seriously ill, but he made notes on logarithms, and almost certainly recognized their relevance to much of his own earlier work.

One cannot help thinking that with Harriot, as with Napier, Briggs could have engaged in lengthy and fruitful conversations. It could so easily have happened because for the last 20 years of Harriot's life they lived not far distant from each other: Harriot at Syon House, Briggs close to Bishopsgate, only a mile from the Tower of London where Harriot regularly visited Raleigh and Percy. There is no evidence, however, that their lives ever coincided. Their circles of friends and spheres of influence were quite different: Briggs was employed by a public institution, while Harriot worked privately from his own home. A treatise by Briggs on 'The northwest passage to the South Sea through the continent of Virginia', which would surely have interested Harriot, was published in 1622, the year after Harriot's death, and Briggs's *Arithmetica logarithmica* not until 1624. During the 1620s, Briggs did come into contact with Harriot's friend Nathaniel Torporley and was aware of attempts to publish some of Harriot's papers, but he himself died in 1630, the year before the posthumous publication of Harriot's *Praxis*. Thus in print, as in life, they sailed close but managed to miss each other.

The lives of Harriot and Briggs offer a pertinent contrast between the older habits of patronage and the new lives of professional mathematicians, properly paid in return for clear responsibilities, particularly in teaching. The latter, of course, was the way of the future.

Institutions, publications, and conferences

The life of Joseph-Louis Lagrange, one of the finest mathematicians of the 18th century, epitomizes some of the new possibilities opening up to a talented mathematician in western Europe 150 years after the deaths of Briggs and Harriot. Lagrange was born in 1736 into a French-Italian family in Turin (his baptismal name was Giuseppe Lodovico Lagrangia). At the age of 17 he discovered a predilection for mathematics and two years later was appointed to teach at Turin's Royal Artillery School. Lagrange was still living with his family in his home town, but intellectually he had already begun to move further afield. Shortly before he took up his teaching post, Lagrange had sent some of his work to Leonhard Euler, director of mathematics at the Royal Academy of Sciences in Berlin. Further letters to Euler rapidly resulted in Lagrange's election to foreign membership of the Academy. At the same time, he and others founded their own scientific society in Turin, one of many such societies founded in western European cities during the 1750s and the forerunner of the present Academy of Sciences of Turin.

The rise of scientific societies and academies is one of the defining features of the intellectual history of the 18th century. The Royal Society of London had been founded in 1660 and the Academy of Sciences of Paris in 1699; a Prussian Academy of Sciences followed in 1700, restructured as the Royal Academy of Sciences of Berlin in 1740, while the St Petersburg Academy of Sciences was founded on the Parisian model in 1724. These institutions offered employment to a small number of mathematicians and scientists; more importantly, their regular meetings provided a forum for the presentation and discussion of new research. Papers presented to such meetings were later published in the academy's *Acta* or *Mémoires*; this process could take some time, but the resulting volumes eventually circulated to readers throughout Europe, and numerous important exchanges were carried out through the pages of academy journals. Lagrange published most of his own early research in the *Mélanges de Turin*, the journal of his own society in Turin.

The Paris Academy also established a tradition of prize questions, with a period of two years for the response. Lagrange sent in entries for the prizes of 1764 (on why the Moon always shows the same face) and 1765 (which he won, on the movement of the satellites of Jupiter). By this time, therefore, he was becoming known and respected by the leading mathematicians of Europe. Jean le Rond d'Alembert, for example, who had earlier been the scientific editor of the *Encyclopédie*, tried hard to find him a post beyond Turin. In 1766, Euler left Berlin for the Academy of St Petersburg and offered to secure a place for Lagrange in Russia, but Lagrange settled instead for Euler's old position at the Academy of Berlin.

The long relationship between Euler and Lagrange, begun before Lagrange was 20, thus remained both intimate and distant. Euler, the most prolific mathematician of the 18th century, threw out one brilliantly intuitive idea after another, but did not always linger long enough to work them through before moving on to the next thing that caught his imagination. The person who very often followed in his wake, turning half-worked ideas into sound and beautiful theories, was Lagrange. Nevertheless, the two never actually met; indeed, Lagrange always retained a respectful distance from Euler, whom he regarded as his elder and superior. He refused to compete directly with Euler for the Paris prize of 1768 (on the motion of the Moon), though they eventually shared the prize of 1772 on a similar subject. Lagrange remained in Berlin for 20 years, during which time he published extensively (in French) in the Academy's *Mémoires*.

After the death of Frederick the Great, who had done so much to support the Berlin Academy, Lagrange moved once again, this time to the Paris Academy, where he arrived in 1787. Two years later, every institution in France was thrown into turmoil by the Revolution, but Lagrange somehow managed through those years to keep his head and his reputation. In 1795 the Academy was abolished and replaced by the Institut National; Lagrange was elected chairman of the section dealing with physical and mathematical sciences. At the same time, the Revolution's pressing need for properly trained teachers and engineers led to the founding of new institutions, in particular the École Polytechnique in 1794 and the École Normale for the training of teachers in 1795. Lagrange taught at both. The École Polytechnique was to become the most prestigious educational institution of early 19th-century Paris. Anyone who has studied mathematics beyond school level will almost certainly be familiar with the names of Lagrange, Laplace, Legendre, Lacroix, Fourier, Ampère, Poisson, and Cauchy, all of whom taught or examined at the École Polytechnique in its early years. Further, the École published its lecture notes in 'cahiers' which were used as textbooks throughout France, especially by those aspiring to be accepted as students.

Lagrange died in 1813. In the first two-thirds of his career, in Turin and Berlin, he had both contributed to and benefited from national academies and their respective journals, institutions which had done much to foster the creation and dissemination of new research. During his final years in Paris, Lagrange saw the rise of a new kind of institution, designed to offer a high level of mathematical and scientific training to the most able students. Unlike the universities, the École Polytechnique offered an education that was tightly focused and practical, and would enable its graduates to consolidate the gains of the Revolution and later the Napoleonic empire.

In case a history of institutions seems somewhat impersonal, let us not lose sight of the close personal relationships that also ran through Lagrange's life, notably with Euler and d'Alembert. When Lagrange died, his protégé, Augustin-Louis Cauchy, son of a family friend, was just embarking on his own long career and was to be a leading figure in French mathematics until his death in 1857. It is possible to trace unbroken chains of personal friendships and collaborations in western European mathematics, from Leibniz in the late 17th century through the Bernoulli family and Euler to Lagrange and on to Cauchy in the mid-19th century.

By the time of Lagrange's death, changes were under way in his earlier home, Berlin. The University of Berlin was founded in 1810 by Wilhelm von Humboldt, as an institution that would not simply pass on accumulated knowledge but which would encourage and facilitate new research. German university professors were free to make their own appointments and so to determine the direction and emphasis of their departments. Research groups, seminars, and doctoral training were all established in the German universities before 1900 and are now imitated, more or less, in every university in the world. Academic mathematicians, Andrew Wiles among them, are all in that sense a product of 19th-century Germany.

The publication of mathematical research changed too. In the 17th and 18th centuries, the main outlets for mathematical articles had been academy journals. The first printed mathematical paper appeared in the *Philosophical Transactions of the Royal Society* in 1668, written by the then president of the Society, William Brouncker. That paper was only four pages long and was juxtaposed with letters to the editor on 'Chymical, Medicinal and Anatomical particulars', on 'the Variety of the Annual High-Tides', and some miscellaneous notices of new books. Journals later became rather better organized: the *Acta eruditorum*, for instance, had separate sections on medicine, mathematics, natural philosophy, law, history, geography, and theology; but scientific journals throughout the 18th century continued to publish on a wide range of subjects of which mathematics was just one.

The first journal dedicated to mathematics alone, the *Annales de mathématiques pures et appliquées*, was founded and edited by

Joseph Gergonne in France in 1810, and became known as Gergonne's journal. Note here the first appearance of a distinction that had not up to then existed in any formal sense between 'pure' and 'applied' mathematics. Gergonne's journal lasted only until 1832, but by then its German equivalent, with a parallel title, had been established in 1826 by August Crelle. The *Journal für die reine und angewandte Mathematik* (Crelle's journal) exists to this day. So does the replacement of Gergonne's journal, first edited by Joseph Liouville in 1836, the *Journal de mathématiques pures et appliquées* (Liouville's journal). Mathematical journal publishing has continued to flourish and increase ever since: today, journals no longer specialize in mathematics as a whole but in the branches and twigs of the discipline. One of the titles I like is the *Journal of Ill-Posed and Inverse Problems*, but there are hundreds of others.

Specialized institutions, entrance examinations, lengthy training, dedicated journals, professional societies, and regular meetings and conferences, are the hallmarks of every modern profession, including mathematics. International or even national conferences did not exist in Lagrange's day, but they certainly do now, and absorb at least some of the time of all academic mathematicians. In particular, mathematicians are always ready to celebrate each other's important birthdays, another sign of the strong social cohesion of the discipline.

The first International Congress of Mathematicians was held in Zürich in 1897, and was attended by representatives of several European countries and the United States. The second Congress, held in Paris in 1900 to coincide with the Exposition Universelle, is best remembered for a speech made by the German mathematician David Hilbert, in which he outlined 23 problems that he hoped mathematicians would solve in the new century (though proving Fermat's Last Theorem was not one of them). After 1900, Congresses were held every four years except during World Wars I and II. The exclusion of mathematicians from Germany, Austria, Hungary, Turkey, and Bulgaria during the

1920s, however, and the absence of others who objected to that ruling, led to debate as to whether these congresses could be called 'international'.

A list of cities that have hosted the Congress tells its own story of the increasingly global nature of mathematical research. Until the 1960s, all the meetings were held in western Europe, Canada, or the United States, but the Congress for 1966 was held in Moscow and for 1982 in Warsaw. The first Asian country to host it was Japan, in 1990, followed by China in 2002 and India in 2010. When Wiles announced his proof of Fermat's Last Theorem in his home city of Cambridge, he could just as easily have spoken to similar audiences in Beijing, Madrid, or Hyderabad, the venues of the three most recent Congresses. Mathematics is now not only a highly professionalized discipline but a thoroughly international one.

By now we have reached the top of the mathematical pyramid, the tightly knit community of professionals that has become associated with the words 'mathematics' and 'mathematician'. Compared to the number of people from schoolchildren upwards who regularly practise mathematics, however, this professional community is tiny, and the number of women in it is even tinier. One can wonder why women are still so under-represented. There is no easy answer to this question, but we should recall that as in most professional spheres the rules were devised by and for men, and it may be that some women find the air at the top of the pyramid a little rarefied and the company not always congenial. If we leave elite mathematics to elite historians, this hardly matters. Just as mathematics itself has gone through many manifestations, so mathematical lives have been lived in a multitude of ways, none more valid or correct than any other.

Chapter 6
Getting inside mathematics

So far, I have avoided too much discussion of mathematical technicalities, and we will not go far into them in this chapter either, but a historian of mathematics is bound to engage not only with the social context of mathematical texts written in the past but also as closely as possible with their content. This is more easily said than done. At one level, the mathematics of the past can seem easy compared with what is expected of, say, a college student today. The difficulty for the historian is usually not so much understanding the mathematics itself as entering into the mind and mathematical universe of someone from another era.

As an example, let us think for a moment about Pythagoras' Theorem, which has now been mentioned several times in this book. Euclid's proof of the theorem is illustrated in Figure 9. It entails drawing the squares on the three sides of the right-angled triangle, dividing the largest one into two parts, and then showing that each of those parts is equal to one of the two smaller squares. The details were cleverly shown in colour by Oliver Byrne in 1847, in the almost wordless proof shown in Figure 10. One of the key features of this proof is that it applies to any right-angled triangle, however you happen to draw it (indeed, David Joyce's interactive version will allow you to push and pull the original triangle around as much as you like as long as you keep the right angle). In other words, the proof does not depend on particular measurements;

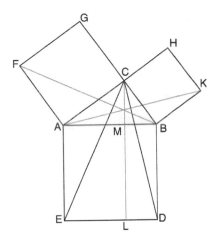

9. Euclid's proof of Pythagoras' Theorem: AFGC = AMLE and CHKB = BDLM

there is no arithmetic involved, and certainly no algebra. This is completely in keeping with the style of the *Elements*: Euclid allowed his readers a straight edge and a compass but no calculator.

Thābit ibn Qurra's proof, shown in Figure 8, also relies on cut-and-paste geometry to demonstrate that the larger square can be made to cover the two smaller ones. For Euclid and ibn Qurra, the underlying intuition behind both the theorem and its proof was geometric.

Now consider the modern technique of labelling the sides of the triangle a, b, and c and writing down $a^2 = b^2 + c^2$. Does this represent the theorem that Euclid had in mind? In one sense, yes. We know that the area of a square with side a is a^2, so the formula is just a very concise way of encapsulating a geometric fact. There is even continuity in the language: we use the same word 'square' for the quantity a^2 and for the four-sided geometric shape. But in another sense, no. The formula comes from a mathematical

91

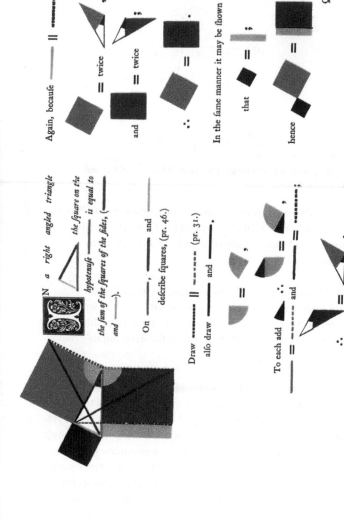

10. Oliver Byrne's proof of Pythagoras' Theorem

culture very different from Euclid's, in which we have learned to let letters represent lengths, and in which we can even forget about geometry and manipulate the letters according to their own rules. Thus, if we wish, we can re-write the above formula as $c^2 = a^2 - b^2 = (a - b)(a + b)$, which is true but no longer has any obvious relevance to a right-angled triangle.

The change from geometric insight to algebraic manipulation is not a trivial one: it takes some effort to learn how to do it. Historically, the shift from a mathematical culture in which geometry was dominant to one in which the language of algebra began to take precedence came about in western Europe in the 17th century. (Fermat was one of the earliest mathematicians to experiment with the possibility, though later he also complained bitterly about any departure from traditional ways of doing things.) Historians have studied this period intensively because the changes were crucial to the development of modern mathematics. To regard the algebraic version of Pythagoras' Theorem as essentially the same as the geometric version is to ignore the historical gulf between them, a gulf that has been crossed only by the cumulative endeavours of many individual thinkers.

Reinterpretation

The example we just looked at is a case of mathematical reinterpretation, in this instance taking a geometric theorem and reinterpreting it algebraically. This is something that mathematicians do a lot; indeed taking an earlier piece of work, their own or someone else's, exploring it, extending it, trying it out under new conditions, is one of the chief ways that mathematicians develop their subject. It is one thing for mathematicians themselves to re-write old mathematics, however, and quite another for historians to do it. When the *Arithmetica* of Diophantus was rediscovered in Europe during the

Renaissance, it turned out to be such a rich source of problems that it was reinterpreted in several ways, both mathematically and historically. We will look at some of the mathematics first.

We have already seen how Fermat extended Problem II.8, testing it out on cubes or even higher powers as well as squares. Here we will look at another early 17th-century reinterpretation, of a different problem from Diophantus, this time by the English mathematician John Pell. Born in Southwick in Sussex in 1611, Pell lived at the same time as both Harriot and Oughtred (whom we met in Chapter 5), though he was some 50 years younger. He was educated at the newly founded Steyning Grammar School, a few miles north of Southwick, and then at Trinity College, Cambridge. Afterwards he returned to Sussex and taught at an experimental school in Chichester until it closed a few years later. Pell then spent several years searching for either a paid post or a patron but found neither that suited his rather particular temperament. At the end of 1643 he was appointed to the Gymnasium in Amsterdam, and two years later to the Illustre School in Breda where he remained until 1652.

During this time Pell gave a good deal of attention to Diophantus. We know this because by the early 1640s Pell had become acquainted with Sir Charles Cavendish (whom we also met in Chapter 5) and they corresponded throughout Pell's years in the Netherlands. They had their own version of a mathematical patronage relationship: Cavendish would ask Pell to help him with whatever he had failed to understand in his latest effort at mathematical reading and Pell would duly respond. Cavendish clearly thought highly of Pell's abilities and expected him to publish a number of important books, including a new edition of Diophantus 'which', he wrote, 'I am exceeding greedie to see'. Unfortunately, Pell was almost pathologically incapable of finishing or publishing anything, but there is evidence that he at least began to work on such an edition.

That evidence comes from Pell's voluminous notes (thousands of pages now bound in more than 30 large volumes in the British Library). The reason that Pell was so interested in Diophantus was that he, Pell, had developed a method of problem-solving that he thought perfectly suited to the questions in the *Arithmetica*. The method was this: first, for any question set down the unknown quantities and the given conditions in numbered lines; second, work systematically from the conditions to the required answer. To ensure that the work proceeds properly, it is set out in three columns, with line numbers in the narrow central column. For each line, the left-hand column contains a brief instruction; the right-hand column shows the result of carrying it out. The whole thing has much the look and feel of a modern computer algorithm.

To see how such a method can be applied to the ancient work of Diophantus, let us look at Pell's rendition of Problem IV.1 of the *Arithmetica*: to find two numbers whose sum is a given number and whose cubes sum to another given number. Diophantus suggested that the sum of the two numbers should be 10 and the sum of their cubes 370, precisely the problem later worked by the young Anne Davenant under her father's instruction. Pell solved it in his own unique style. His first two lines are the following, in which he calls the unknown numbers a and b.

$$a = ? \quad \Big|\ 1\ \Big|\ aaa + bbb = 370$$
$$b = ? \quad \Big|\ 2\ \Big|\ a + b = 10$$

Next, following Diophantus exactly, Pell introduced a third number, c, and supposed that $a = c + 5$ so that, necessarily, $b = 5 - c$. His next two lines are therefore

$$c = ? \quad \Big|\ 3\ \Big|\ \text{let } a = c + 5$$
$$2' - 3' \quad \Big|\ 4\ \Big|\ b = 5 - c$$

where $2' - 3'$ just means subtract line 3 from line 2. Now everything is set up and the work can proceed. The reader who wants to follow the detail needs to know that Pell's instruction $3' @ 3$ means take the cube of line 3, while $10' \omega 2$ means take the square root of line 10. A further convention favoured by Pell is switching from lower- to upper-case letters once the required values have been found.

$3'@3$	5	$aaa = ccc + 15cc + 75c + 125$
$4'@3$	6	$bbb = 125 - 75c + 15cc - ccc$
$5' + 6'$	7	$aaa + bbb = 30cc + 250$
$7', 1'$	8	$30cc + 250 = 370$
$8' - 250$	9	$30cc = 120$
$9' \div 30$	10	$cc = 4$
$10'\omega2$	11	$c = 2$
$11' + 5$	12	$c + 5 = 7$
$5 - 11'$	13	$5 - c = 3$
$3', 12'$	14	$A = 7$
$4', 13'$	15	$B = 3$

The final four lines check that the problem has indeed been correctly solved:

$14'@3$	16	$AAA = 343$
$15'@3$	17	$BBB = 27$
$16' + 17'$	18	$AAA + BBB = 370$
$14' + 15'$	19	$A + B = 10$

It seems that Pell planned to re-write the whole six books of the *Arithmetica* in this style, but if he ever completed them his manuscript is lost. Many of his contemporaries were impressed with his method, though. His friend John Aubrey even invented a new Latin verb for it: *pelliare*, 'to pelliate'.

It is clear from the above example that Pell did not believe in wasting words: the only one that appears in his 19 lines of working is 'let' (he actually wrote it in Latin as *sit*). But if words are to disappear, there must be symbols to replace them, and here Pell was a master of invention. The symbols @ and ω that helped to keep his left-hand column concise have long since fallen out of use, but his division symbol $\div$ is still with us. Invention of notation was one of Pell's particular talents; in this he was following what was something of an English tradition at the time. By 1557 Robert Recorde had devised the = sign based on two parallel lines 'bicause noe 2 thynges, can be moare equalle'. Around 1600 Thomas Harriot had added the inequality signs < and > and the convention of writing ab for a multiplied by b. In 1631 William Oughtred introduced the × sign, though he rarely used it; he also argued passionately that notation 'plainly presenteth to the eye the whole course and processe of every operation and argumentation'. This was clearly what Pell thought too, that his method rendered the argument plain to the eye without need for further explanation. His efforts to 'pelliate' Diophantus therefore tell us rather more about the aspirations of early 17th-century English algebraists than they do about Diophantus and his *Arithmetica*.

This is true of reinterpretations of a historical rather than mathematical kind too: that they generally reveal more about the interpreter than the interpreted. Stories that have circulated over the centuries about the origins of algebra, for example, have recorded not just historical fact but contemporary understanding. Algebra first came to non-Islamic regions of western Europe in the late 12th century through translations of al-Khwārizmī's *Al-jabr wa'l muqābalā*, but by the 16th century, this early history had been forgotten, if indeed it had ever been properly known. Nevertheless, the Islamic origins of the subject were recognized, if only from the strange-sounding words 'algebra' and 'almucabala' associated with it. Thus early 16th-century writers ascribed the invention of algebra variously to '*une nome Arabo di grande intelligentia*' ('a certain Arab of great intelligence'), sometimes to one Algeber (actually Jābir ibn Aflaḥ, a 12th-century Spanish Muslim astronomer who had nothing to do with it), or to the vaguely named 'Maumetto di Mose Arabo' (a rendering of Muḥammed ibn Mūsā, an Arab).

In 1462, however, the German scholar Johannes Müller, usually known as Regiomontanus after the Latinized name of his home town, Königsberg, examined a manuscript of Diophantus'

Arithmetica in Venice. Three years later, lecturing in Padua, he described the contents as 'the flower of all arithmetic ... which today is called by the Arabic name of algebra'. His lecture was not published until 1537, but very soon after that other writers began to pick up the same theme: that algebra had been invented by Diophantus and was only later adopted by 'the Arabs'. One can see why such stories were accepted at a time when a Greek pedigree conferred instant respectability and status. The fact that the problems handled by Diophantus were different in both style and content from those found in Islamic texts does not seem to have deterred anyone from thinking that the latter must somehow have been derived from the former.

Even today, when there is much greater appreciation of the mathematics that western Europe inherited from the Islamic world, Diophantus is still sometimes credited as being the founder of algebra. This is a debate that can run and run, but we should try to understand mathematically what is at stake. It is true that Diophantus posed several 'find a number' problems that are easily handled by modern algebraic methods, as Pell's example demonstrates. But he also posed a great many other problems that are 'indeterminate', that is, they have more than one possible answer. In such cases, Diophantus was usually satisfied if, by some special method, he could show just one of those answers. In fact his work is full of ideas, some of them very clever, that work for specific questions, unlike the more general rules of the later Islamic algebra texts. It has also been suggested that Diophantus used an elementary symbolic notation by writing, for example, ς for an unknown number and Δ^Υ for its square, but it has now been shown that these abbreviations for the Greek words *arithmos* (number) and *dynamis* (square), respectively, were introduced by 9th-century copyists and cannot be attributed to Diophantus at all. Finally, the mathematics that has been derived from the *Arithmetica* has been absorbed into modern number theory, whereas Islamic *al-jabr* texts gave rise much more directly to the algebra of western Europe. It seems to me that the word 'algebra' should be reserved for the rules and procedures that were described by the participants themselves as 'al-jabr' or 'algebra', and that we should not impose those words, nor the history they carry with them, on a writer who was working in an earlier and very different tradition.

Who was first . . . ?

The question we have just examined, 'Who invented algebra?', is typical of those sometimes asked of historians of mathematics, who are often expected to be able to say who was first to discover or invent certain ideas. Except in the simplest cases such questions can be extraordinarily difficult to answer. Take, for example, the discovery of calculus. This is the branch of mathematics that can be used for describing and predicting change. It is used today in biology, medicine, economics, ecology, meteorology, and every other science that works with complex interactive systems. It is therefore not unreasonable to want to know 'Who invented calculus?'

The short answer is that two people did, almost simultaneously but independently: Isaac Newton working in Cambridge and Gottfried Wilhelm Leibniz working in Paris. To modern historians, there is no longer any dispute about this because we have the manuscripts of both men and can see exactly when and in what order their ideas were developed. We can also see that they approached the work in very different ways, each devising their own vocabulary and notation (Leibniz spoke of 'differentials' while Newton spoke of 'fluxions'; Leibniz invented the now familiar $\frac{dx}{dt}$ notation whereas Newton used the now less common $\dot{x}$).

For their contemporaries, however, the story was not clear at all. The basic facts are that Newton developed his version of calculus during 1664 and 1665 (before his 23rd birthday) but then did nothing with it. By the early 1670s he had already engaged in an intellectual skirmish with Robert Hooke over his optical findings and was perhaps reluctant to risk another over the calculus. In any case, by that time his interest had shifted to alchemy, which was to preoccupy him for the next decade. In 1673, however, Leibniz, then living in Paris, independently began to work on some of the same problems that had earlier intrigued Newton, and published his first paper on calculus in 1684 followed by others in the 1690s. Newton appears to have taken little notice, probably regarding Leibniz's early work as rather trivial compared with what he himself had been able to achieve. Some of Newton's friends felt differently, however, and around the turn of the century his English supporters began to hint not only that Newton had been first but that Leibniz might actually have stolen the seeds of his ideas from Newton. It did not help Leibniz's case that he had seen some of Newton's papers when he was in London in 1675 and had received letters from Newton in 1676, but what he had learned from them, and how that related to what he had discovered already, no-one but Leibniz really knew.

Both Newton and Leibniz held back from direct confrontation but allowed the battle to be fought out through their henchmen, who were thoroughly belligerent on both sides. Eventually, in 1711, Leibniz appealed to the Royal Society, of which he was a member, to adjudicate in the dispute. Newton, as President of the Society, set up a committee which barely needed to meet because Newton was already busy writing its report. Not surprisingly, it found in Newton's favour. And, also not surprisingly, that was not the end of the matter: the dispute rumbled on until after Leibniz's death in 1716. The dispute explains why in 1809 the English schoolboy George Peat in Cumbria learned a subject called 'fluxions' rather than a subject called 'calculus'.

It is an unedifying story from which no-one comes out well. The point of re-telling it is to emphasize how difficult it was for anyone at the time to get to the bottom of it: no single person was in possession of all the facts; besides, it was difficult to know whether the argument was about the calculus as a whole or about particular aspects of it (Leibniz accused the English of shifting ground on this); and, as can be the way with disputes, several disagreements were dragged in that were never part of the original argument. Another point of the story, however, is that the ultimate evidence for the truth comes not from what people at the time wrote or said, which was almost always partial (in both senses of the word), but from the mathematical manuscripts themselves.

In mathematics it is not at all uncommon, as in the case of the calculus, for two people to come up with similar ideas at more or less the same time. Once the groundwork has been laid, one mathematician can make use of it just as easily as another, and it then becomes very difficult to apportion credit, especially if the two have had some contact with each other. It was for precisely this reason that Wiles shut himself away so carefully during his years of work on Fermat's Last Theorem. In the case of the calculus, there is enough documentary evidence for historians to work out what really happened but this is not always so. Two early 19th-century mathematicians, Bernard Bolzano in Prague and Augustin-Louis Cauchy in Paris, also developed some remarkably similar ideas, Bolzano in 1817, Cauchy in 1821. Did Cauchy 'borrow' from Bolzano or not? Bolzano's work was published in a little-known Bohemian journal which was nevertheless available to Cauchy in Paris. On the other hand, both could have built independently on the earlier work of Lagrange. We might also throw into the assessment circumstantial evidence about Cauchy's way of working, which was very often to pick up good ideas from someone else and develop them at length. In the end, for lack of firm evidence either way, we simply cannot say.

Another problem about saying who was first to make a discovery can be defining what we think the discovery actually consists of. At what precise point in history, for example, can we say we have 'calculus', as opposed to a tangle of related ideas that gradually began to make sense first to Newton and later to Leibniz? It is just as difficult, as we have already seen, to pinpoint where algebra began, or where Pythagoras' Theorem became a formal theorem as opposed to a useful fact known to builders. Almost all new mathematics is built on previous work, and sometimes on a number of contributory ideas. Tracing the antecedents of a particular technique or theorem is one of the tasks of the historian, not in order to say who was first, but to understand more clearly how mathematical ideas have changed over time.

Getting things right

Euclid's systematic deductive style, in which every theorem is carefully proved from the theorems and definitions that have gone before it, has stood for centuries as the gold standard of mathematical style. But even Euclid turned out not to be infallible. Questions were raised about one of Euclid's postulates as early as the 5th century AD and proved very difficult to answer. The troublesome postulate is sometimes known as the Parallel Postulate; it can be expressed in different ways, but the simplest is to say that if we have a line l in a plane, and a point P not on the line, there is just one line through P parallel to l. Most of us would have no difficulty accepting that. It leads to the consequence that the angles in a triangle add up to 180 degrees, and most of us have no difficulty accepting that either. Many commentators on Euclid, however, thought that the Parallel Postulate should not be a postulate but a theorem, that is, that it should somehow be possible to prove it from the other definitions and postulates. Thābit ibn Qurra and Umar al-Khayyāmī were amongst those who tried; so was John Wallis in Oxford in 1663. Then in 1733 an otherwise little remembered mathematician called Gerolamo Saccheri, professor of mathematics at Pavia in northern Italy, tried a different approach. He investigated what would happen if he assumed that the angles of a triangle add up to either less or more than 180 degrees, hoping, of course, that the results would be absurd so that such assumptions could be dismissed. He was wrong. Assuming that the angles add up to less than 180 degrees led him to some strange but nevertheless consistent results.

A hundred years later, Nicolai Ivanovich Lobachevskii, professor at the university of Kazan in Russia, and János Bolyai from the town that is now Cluj in northern Romania, took these ideas much further (another example of independent but more or less simultaneous discovery); they both realized that it is possible to construct a kind of geometry that is mathematically acceptable but definitely non-Euclidean. The idea was startling to 19th-century thinkers: one consequence was that no-one could know whether infinite space itself was Euclidean or non-Euclidean, any more than we can tell from walking down the street whether the Earth is round or flat. Mathematics was supposed to offer indisputable truths about the world, but suddenly such truths had come to seem less secure.

One of the outcomes of all this was that mathematicians began to look more carefully at their underlying assumptions, formally known as axioms. Indeed, in the late 19th and early 20th centuries, in a return to true Euclidean style, entire branches of mathematics came to be set up on axiomatic foundations, imposing on them a logical rigour that mathematics had not known since the Greek era. For between the 2nd century BC and the 19th century AD, mathematics had developed for the most part in a thoroughly haphazard way. The fact is that mathematicians do not make discoveries by setting up axioms and thinking logically about them but by responding imaginatively to problems that interest them, by opening up questions in new directions, or by seeing how apparently different bits of mathematics might fit together in a fresh way. Of course, they must apply their skills and experience correctly, and in the end they must present a watertight argument known as a 'proof', as Wiles did in his Cambridge lectures, but that is likely to be some way down the road from the initial insights and the hard work that almost invariably follows them.

The discovery of the calculus, discussed in the previous section, is a supreme example of mathematics that in its inception was not logical at all. The whole idea was based on what 17th-century mathematicians called 'infinitely small quantities'. The question one is forced to ask about an infinitely small quantity, however, is: does it have any size at all? If it does, then it cannot be 'infinitely small'; but if it has no size, then it does not even exist and one cannot use it in any sensible way in one's calculations. This may seem like nit-picking, more like a discussion of angels on a pinhead than of mathematics. But it matters, because discussions of infinitely small quantities can rapidly lead to contradictions; and since mathematics is supposed to be a unified logical edifice, a single contradiction brings the whole lot tumbling down. (For this reason mathematicians often deliberately set up a contradiction, as Saccheri tried to do, if they want to prove that something is impossible; the technique is called *reductio ad absurdum*.)

Both Newton and Leibniz were well aware of the paradox of the infinitely small and did their best to deal with it, Newton by addressing it head on, Leibniz by skirting round it. Those who came after them were aware of it too, not only mathematicians but well-educated members of the public also. Bishop George Berkeley, for instance, in a book called *The Analyst: A Discourse Addressed to an Infidel Mathematician* asked 'whether mathematicians, who are so delicate in religious points, are strictly scrupulous in their own science? Whether they do not submit to authority, take things upon trust, and believe points inconceivable?' Did such concerns stop mathematicians in their tracks? No, because very early in the evolution of the calculus they had recognized how powerful it could be and were busy applying it, with plenty of success, to rays of light, hanging chains, falling bodies, vibrating strings, and many other phenomena of the physical world. They were hardly going to give up all this for what they regarded as a metaphysical rather than mathematical difficulty. It took about 150 years until the problem was resolved to most people's satisfaction, in ways that are a little too technical to discuss here; during those same 150 years, however, mathematics advanced beyond all expectations, despite the shakiness in its foundations.

A similar story can be told for the 19th century. In 1822, Joseph Fourier, a lecturer at the École Polytechnique in Paris, published a treatise on the diffusion of heat, his *Théorie analytique de la chaleur*. In it, Fourier investigated the idea of using infinite sums of sines and cosines to describe periodic distributions; these infinite sums are now known as Fourier series, and have a vast range of applications in engineering and physics. Fourier's original derivation, however, was riddled with errors and inconsistencies. Some of these cancelled each other out, but many of them Fourier ignored, if indeed he ever noticed them. In other words, the initial theory of Fourier series was no more firmly grounded than the calculus had been, and yet, like the calculus, it proved to be an immensely rich and useful tool. But just as with the calculus, many mathematicians after Fourier had to spend a great deal of time mending the holes.

These examples are not exceptional. As we have seen, Wiles, a far more competent mathematician than Fourier, had to go through a very similar process of fixing an error, though in his case it took only two years and not a century. Almost every new discovery in mathematics starts out in a rough and ready state and has to be improved and refined before it can be presented to one's peers, let alone taught to beginners.

Most modern textbooks follow the same pattern as Euclid's *Elements*, starting from simple beginnings and building up the mathematics in a seamless logical flow. In other words, we allow – or expect – students to follow a clear-cut path that the first explorers could not even see. If students are given the opportunity to go back to the original discoveries, they are likely to find something very different: a process of trial and error, false starts, and dead ends; half-formed ideas, half worked out, left to be developed by someone else; better-formed ideas refined over months or years; all of it finally adapted by teachers who may not be innovators but who must have the equally important gift of seeing how to explain things to novices. In other words, the polished exposition of a textbook tells us very little about the intuition and hard work, the imagination and struggle, that went into the mathematics in the first place. That is the task of the historian.

Chapter 7
The evolving historiography of mathematics

Ways of doing and thinking about the history of mathematics have changed a great deal over the last few centuries, some of them in keeping with changes in intellectual history more generally, some of them peculiar to mathematics. As we saw in Chapter 2, the approach taken by John Leland in the 1550s, followed by Johann Gerard Vossius a century later, was to record as many facts as possible about authors, dates, and texts, but without any analysis of what those texts contained. By the late 17th century, however, it was clear to anyone interested in mathematics that the power, scope, and techniques of the subject were advancing rapidly: 'Geometry is improving daily', wrote Joseph Glanville in 1668, while just a few years later John Wallis extolled the 'progress and advancement' that had brought algebra to 'the heighth at which now it is'.

The 18th century, the age of the *Encyclopédie*, saw two substantial publications concerned with the history of mathematics, Jean-Étienne Montucla's *Histoire des mathématiques* published in Paris in 1758 (expanded to four volumes in 1799–1802), and Charles Hutton's *Mathematical and Philosophical Dictionary*, which included a number of historical articles, published in London in 1795. By the late 19th century, however, the focus was shifting, as in other areas of study, away from second-hand accounts to scholarly editions and translations of ancient and medieval texts (as had also happened during the Renaissance). To take examples from texts that have been discussed earlier in this book: the first English rendering of the *Arithmetica* of Diophantus was published by Thomas Heath in 1885; Heath's edition of Euclid's *Elements*, based on the best scholarship then available, appeared in 1908; Charles Louis Karpinski's translation of al-Khwārizmī's *Al-jabr* from a medieval Latin version appeared just a few years later, in 1915. Such editions were and remain invaluable: neither the *Arithmetica* nor the *Al-jabr* had previously been available in English; Heath's *Elements* remains the standard English edition to this day.

Modern historians, however, also treat such editions with some care. Heath's essay on the *Arithmetica* is entitled *Diophantus of Alexandria: A Study in the History of Greek Algebra*, a title that raises questions that have already been addressed in this book. Further, a commentator on Heath's edition of Apollonius has observed that 'thanks to skilful compression and the substitution of modern notation for literary proofs, [it] occupied less than half the space of the original'. Again, the historian might not thank Heath for this skill, preferring to see the text uncompressed and free of the anachronisms of modern notation. A good deal of early 20th-century scholarship in the history of mathematics, however, often carried out by mathematicians rather than historians, proceeded in much the same way, translating texts originally written in Egyptian hieroglyphics or in Sumerian, Sanskrit, or Greek into the symbols and concepts of modern mathematics. The motives of the translators were not in themselves reprehensible: in trying to understand ideas that at first seem thoroughly alien, it is natural to try to relate them to something more familiar; the danger is that one comes to regard unfamiliar ideas as no more than archaic renderings of what we can now do, as we see it, more efficiently. In this way, history comes to be re-written from our own perspectives instead of those of the original authors.

Historians of ancient mathematics were among the first to rebel against the distortions introduced by modernization, and during the 1990s led the way in attempting to recover and preserve, as far as possible, the idiom and thought processes of the originals. As Reviel Netz, editor and translator of Archimedes, has said in a remark now often quoted: 'the purpose of a scholarly translation as I understand it is to remove all barriers having to do with the foreign language itself, leaving all other barriers intact'. This forces the modern reader of historical mathematical texts to work much harder than the reader of 50 years ago, but the gains in historical understanding are incomparably greater.

Those who work with ancient mathematical writings have led the way in other aspects of historiography too, partly because of the way their material has been so haphazardly accumulated in the past. A single clay tablet, for instance, tells us very little unless we know where or when it was written. Such information is essential if we are to build up a picture of how a particular text relates to others found in the same area or elsewhere. Many tablets from early excavations were deposited in museums with minimal information on their provenance, or sold on the antiquities market with none, making it exceedingly difficult for historians to deduce useful information from them now. Fortunately, present-day archaeologists record positions and surroundings with great care before removing each layer of evidence. Modern technology has also enabled advances in reading faded or damaged pen and ink texts. Work on the rediscovered text by Archimedes mentioned in Chapter 3 has been particularly remarkable in this respect. Scholars have been able not only to read much of the original text but to identify the scribe who scrubbed the parchment clean and overwrote it: Ioannes Myronas, working in Constantinople during Lent 1229. It is entirely appropriate that recovery of the text and recovery of the story of the text should go hand in hand.

Historians of mathematics have increasingly moved away from a purely 'internalist' view in which mathematical developments are seen to come about of their own accord, regardless of outside influences. As has now been shown over and over again in this book, mathematical activity has for centuries manifested itself in a variety of ways, all of them socially and culturally determined. We should not throw out the baby with the bathwater, however: mathematicians often devote themselves to a particular problem not because it might be useful or because anyone requires them to do so, but because the problem itself catches their imagination. This was precisely the case for Newton and Leibniz with the calculus, Bolyai and Lobachevskii with non-Euclidean geometry, or Wiles with Fermat's Last Theorem. In such cases, progress depends first and foremost on deep and concentrated engagement with the mathematics, and in that sense mathematical creativity can be said to be an internal process. But the mathematical questions that are considered important at a particular time or place, the way they have come to be there, the way they are understood and interpreted, are all influenced by a multitude of factors outside the mathematics itself: social, political, economic, and cultural. Context has become as important to the historian as content.

Another significant change in recent years has been the growing recognition that the mathematics done by a small number of famous mathematicians has not reflected (though it has built on) the diversity of mathematical activity and experience at other levels of society. The history of non-elite mathematics has been one of the key themes of this book. Historians of mathematics, like scholars in many other disciplines, have also become much more sensitive to questions of gender and ethnicity. Studies of cultures prior to or beyond modern western Europe have been constrained in the past by lack of sources, or linguistic barriers, but that situation is now beginning to change as web images, new translations, and scholarly commentary are making increasing amounts of source material more readily accessible, intellectually as well as materially. Consequently, the mathematics of the past is no longer regarded simply as a precursor to the mathematics of the present but as an integral part of its own contemporary culture.

As in all thriving academic disciplines these days, those who engage in the history of mathematics are required to cross boundaries. Indeed, one of the great pleasures of working in the subject is that one can learn from the expertise of archaeologists, archivists, sinologists, classicists, orientalists, medievalists, historians of science, linguists, art historians, literary critics, museum curators, and many others. The range of sources has broadened similarly and is no longer restricted to the books or manuscripts that once expounded the latest ideas but includes correspondence, diaries, rough notes, exercise books, measuring instruments, calculating machines, paintings, sketches, diaries, and novels. The last item may sound surprising, but novelists may be the most astute and articulate recorders of contemporary views of mathematics; readers interested in following up this theme will find more under 'Further reading'.

The questions asked by historians over the last 50 years have both changed and diversified. It is no longer enough simply to ask who discovered what and when. We also want to know what mathematical practices engaged groups of people or individuals, and why. What historical or geographical influences were at work? How were mathematical activities perceived, by the participants or by others? What aspects were particularly valued? What steps were taken to preserve or hand on mathematical expertise? Who was paying for it? How did an individual mathematician manage his (or her) time and skill? What were their motivations? What did they produce? What did they do with it? And with whom did they discuss, collaborate, or argue along the way?

Most of the answers to most of these questions will be difficult to find with any degree of certainty. Historians of mathematics, like all other historians, work with sparse evidence, from which they must reconstruct as carefully as possible incomplete stories about the past. The attempt remains worthwhile for what we can learn about a human activity as ancient and as widespread as producing literature or music, and which has manifested itself in as rich a variety of cultural forms: doing and creating mathematics.

Further reading

Chapter 1

Primary sources

Robert Recorde, *The Pathway to Knowledg* (London, 1551);
painstakingly reprinted by Gordon and Elizabeth Roberts (TGR
Renascent Books, 2009).

Secondary sources

Markus Asper, 'The two cultures of mathematics in ancient Greece', in
Eleanor Robson and Jacqueline Stedall (eds), *The Oxford
Handbook of the History of Mathematics* (Oxford University Press,
2009), pp. 107–132.

Simon Singh, *Fermat's Last Theorem* (Fourth Estate, 1997; Harper
Perennial, 2007).

Benjamin Wardhaugh, 'Mathematics in English printed books,
1473–1800: a bibliometric analysis', *Notes and Records of the Royal
Society*, 63(2009): 325–38.

Chapter 2

Primary sources

*The Suàn shù shū, writings on reckoning: a translation of a Chinese
mathematical collection of the second century* BC, *with explanatory
commentary*, tr. Christopher Cullen (Needham Research Institute,
2004).

Fibonacci's Liber abaci: Leonardo Pisano's book of calculation, tr. L. E.
Segal (Springer, 2002).

Secondary sources

Christopher Cullen, 'People and numbers in early imperial China', in
Eleanor Robson and Jacqueline Stedall (eds), *The Oxford
Handbook of the History of Mathematics* (Oxford University Press,
2009), pp. 591–618.

G. E. R. Lloyd, 'What was mathematics in the ancient world?', in
Eleanor Robson and Jacqueline Stedall (eds), *The Oxford
Handbook of the History of Mathematics* (Oxford University Press
2009), pp. 7–25.

Benjamin Wardhaugh, 'Poor Robin and Merry Andrew: mathematical
humour in Restoration England', *BSHM Bulletin*, 22(2007): 151–9.

Chapter 3

Primary sources

The thirteen books of Euclid's *Elements* in MS D'Orville 301, from
888, http://www.claymath.org/library/historical/euclid/ last
accessed November 2011.

David Joyce, 'A Quick Trip through the Elements', compiled in 2002,
http://aleph0.clarku.edu/~djoyce/java/elements/trip.html, last
accessed January 2012.

Secondary sources

June Barrow-Green, ' "Much necessary for all sortes of men": 450 years
of Euclid's *Elements* in English', *BSHM Bulletin*, 21(2006): 1–25.

Annette Imhausen, 'Traditions and myths in the historiography of
Egyptian mathematics', in Eleanor Robson and Jacqueline Stedall
(eds), *The Oxford Handbook of the History of Mathematics* (Oxford
University Press, 2009), pp. 781–800.

Victor Katz (ed.), *The mathematics of Egypt, Mesopotamia, China,
India, and Islam: a sourcebook* (Princeton University Press, 2007).

Reviel Netz and William Noel, *The Archimedes Codex: revealing the
secret of the world's greatest palimpsest* (Weidenfeld and Nicolson,
2007).

Eleanor Robson, *Mathematics in ancient Iraq: a social history*
(Princeton University Press, 2006).

Corinna Rossi, 'Mixing, building, and feeding: mathematics and
technology in ancient Egypt', in Eleanor Robson and Jacqueline

Stedall (eds), *The Oxford Handbook of the History of Mathematics* (Oxford University Press, 2009), pp. 407–28.

Benjamin Wardhaugh, *How to read historical mathematics* (Princeton University Press, 2010).

Chapter 4

Primary sources

Copy books in the John Hersee collection owned by the Mathematical Association, in the David Wilson Library at the University of Leicester.

Secondary sources

Marit Hartveit, 'How Flora got her cap', *BSHM Bulletin*, 24(2009): 147–58.

Eleanor Robson, *Mathematics in ancient Iraq* (Princeton University Press, 2008).

Polly Thanailaki, 'Breaking social barriers: Florentia Fountoukli (1869–1915)', *BSHM Bulletin*, 25(2010): 32–8.

Chapter 5

Sonja Brentjes, 'Patronage of the mathematical sciences in Islamic societies', in Eleanor Robson and Jacqueline Stedall (eds), *The Oxford Handbook of the History of Mathematics* (Oxford University Press, 2009), pp. 301–27.

Chapter 6

Primary sources

Euclid, *The first six books of the Elements of Euclid*, beautifully reproduced in colour from Oliver Byrne's 1847 original by Werner Oechslin and Petra Lamers-Schutze (Taschen, 2010).

Euclid, *Elements*, Oliver Byrne's coloured edition of 1847 combined with David Joyce's interactive version of 2002 http://www.math.ubc.ca/~cass/Euclid/byrne.html, last accessed January 2012.

Secondary sources

Glen van Brummelen, 'Filling in the short blanks: musings on bringing the historiography of mathematics to the classroom', *BSHM Bulletin*, 25(2010): 2–9.

Chapter 7

Tony Mann, 'From Sylvia Plath's *The Bell Jar* to the Bad Sex Award: a partial account of the uses of mathematics in fiction', *BSHM Bulletin*, 25(2010): 58–66.

Index

NUMBERS
A Very Short Introduction
Peter M. Higgins

Numbers are integral to our everyday lives and feature in everything we do. In this *Very Short Introduction* Peter M. Higgins, the renowned mathematics writer unravels the world of numbers; demonstrating its richness, and providing a comprehensive view of the idea of the number. Higgins paints a picture of the number world, considering how the modern number system matured over centuries. Explaining the various number types and showing how they behave, he introduces key concepts such as integers, fractions, real numbers, and imaginary numbers. By approaching the topic in a non-technical way and emphasising the basic principles and interactions of numbers with mathematics and science, Higgins also demonstrates the practical interactions and modern applications, such as encryption of confidential data on the internet.

www.oup.com/vsi

SCIENTIFIC REVOLUTION
A Very Short Introduction
Lawrence M. Principe

In this *Very Short Introduction* Lawrence M. Principe explores the exciting developments in the sciences of the stars (astronomy, astrology, and cosmology), the sciences of earth (geography, geology, hydraulics, pneumatics), the sciences of matter and motion (alchemy, chemistry, kinematics, physics), the sciences of life (medicine, anatomy, biology, zoology), and much more. The story is told from the perspective of the historical characters themselves, emphasizing their background, context, reasoning, and motivations, and dispelling well-worn myths about the history of science.

www.oup.com/vsi

ONLINE CATALOGUE

Very Short Introductions

Our online catalogue is designed to make it easy to find your ideal Very Short Introduction. View the entire collection by subject area, watch author videos, read sample chapters, and download reading guides.

http://fds.oup.com/www.oup.co.uk/general/vsi/index.html